BUFFALO LESSONS

BUFFALO LESSONS

How Bison Returned to Banff National Park

KARSTEN HEUER

Foreword by **LEROY LITTLE BEAR**
Introduction by **HARVEY LOCKE**

DAVID SUZUKI INSTITUTE

GREYSTONE BOOKS
Vancouver/Berkeley/London

Greystone Books Ltd.

greystonebooks.com

David Suzuki Institute

davidsuzukiinstitute.org

Cataloguing data available from Library and Archives Canada

ISBN 978-1-77840-314-9 (pbk.)

ISBN 978-1-77840-315-6 (epub)

Editing by Paula Ayer
Proofreading by Alison Strobel
Cover and interior design by Fiona Siu
Cover image (composite) by Karsten Heuer and DC_Colombia/iStock
Inside cover image by Karsten Heuer
Interior photographs by Karsten Heuer, except for
pages 2 and 10 by Dan Rafla; page 84 © by Parks Canada/Johane Janelle;
and pages 36, 88, 91, 93, and 94 © by Parks Canada/Dan Rafla

Printed and bound in China on FSC® certified paper
at Shenzhen Reliance Printing. The FSC® label means that
materials used for the product have been responsibly sourced.

Greystone Books thanks the Canada Council for the Arts,
the British Columbia Arts Council, the Province of British Columbia
through the Book Publishing Tax Credit, and the Government
of Canada for supporting our publishing activities.

EU Safety Information: Easy Access System Europe, Mustamäe tee 50,
10621 Tallinn, Estonia, gpsr.requests@easproject.com.

MIX
Paper | Supporting
responsible forestry
FSC   FSC® C102842
www.fsc.org

Canadä

BRITISH
COLUMBIA

BRITISH COLUMBIA
ARTS COUNCIL
An agency of the Province of British Columbia

Canada Council   Conseil des arts
for the Arts   du Canada

Greystone Books gratefully acknowledges the xʷməθkʷəy̓əm (Musqueam),
Sḵwx̱wú7mesh (Squamish), and səlilwətaɬ (Tsleil-Waututh) peoples on
whose land our Vancouver head office is located.

For my other teacher, Leanne.

BUFFALO CAME FROM a hole in the ground in numbers so great they would support us forever. But when Europeans came the animals disappeared. They went back in that hole in the ground. The buffalo will eventually return. When they do, it will be from the sky and the mountains.

ELDERS' TELLING OF BLACKFOOT CREATION STORY,
AS RECORDED BY AMETHYST FIRST RIDER

CONTENTS

THE REST
OF THE STORY

Then a Husband
and Wife Came Along...

MILLIONS OF BUFFALO roamed and eco-engineered the plains of North America before and after the last ice age through mutual relationship with the environmental totality. Prophecies by North American Indians such as the Blackfoot, who greatly depended on the Buffalo, forewarned that the Buffalo will disappear if not respected.

It came to be that STRANGE PEOPLE FROM ACROSS THE WATERS APPEARED ON THE SCENE. Through their government and lifeways they brought about near extinction of the Buffalo, fulfilling the prophecy. But the prophecy also told that when the people have learned their lesson, the Buffalo will return from the mountains where they were hidden by NAPI, the Blackfoot trickster.

THEN A HUSBAND AND WIFE CAME ON THE SCENE. The HUSBAND, a wildlife scientist, and his WIFE, a

filmmaker, had followed a caribou herd for five months in the Canadian North. They came to know caribou by "being caribou": just as social scientists studying a culture other than their own, in order to really understand that culture, must come to know everything about the people who live that culture, from world views to customs to social values to the environment where the culture is lived and practised. One can say they were "cariboued."

The HUSBAND and WIFE, having experienced nature trails and the outback of several environmental biomes, decided to study another culture: the Buffalo culture. In their minds, the situation with Buffalo was a bit different: Buffalo in the wild had not existed for nearly 150 years. They initially thought, "We may have to train them to once again live in the wild." But to their continuing surprise, the Buffalo, it seemed, had not lost a step, and went from near extinction to a thriving wild herd in Banff National Park. In the process of rewilding, it turned out the Buffalo

taught the HUSBAND and WIFE all about Buffalo culture and their eco-engineering skills. The HUSBAND and WIFE loved the wonderful lessons taught by the Buffalo so much that one can say they were "buffaloed."

Buffalo Lessons: How Bison Returned to Banff National Park is the fantastic story, and a deep understanding of nature from a Buffalo perspective. The HUSBAND and WIFE, of course, are KARSTEN HEUER and LEANNE ALLISON. The reintroduction of Buffalo into Banff National Park is a remarkable fulfillment of "the rest of the story."

LEROY LITTLE BEAR

INISKIM/UNIVERSITY OF LETHBRIDGE

INTRODUCTION

THERE IS SOMETHING about buffalo that makes them special. This has always been so. To the First Peoples of North America who live in the range of plains bison (which they call "buffalo" in English), these magnificent animals are central to their culture as teacher, provider, and main actor in their stories.

No animal is more symbolic of conservation. The decimation of plains bison is one of the great ecological tragedies of all time. At the beginning of the nineteenth century, millions of these great beasts ranged in enormous herds across the Great Plains and up into mountain landscapes. The arrival of Europeans led to a wanton slaughter. By the late 1880s fewer than a hundred survived. Then followed the successful rescue of plains bison from the brink of extinction, marking the birth of a wider concern for endangered species worldwide.

Questions about what it means to save a species also find their origins with the plains bison. Is it enough to rescue a few animals and keep them alive? Given the similarity between bison and their cattle cousins, should we keep a small number for display and simply accept that they have been replaced by the easier-to-manage cow as part of the march of progress? Or should we treat them like cattle: allow them to exist in large numbers behind fences for eventual slaughter and consumption but treat them as strays if they are not contained? For most of the twentieth century, we answered those questions in the affirmative.

As the turn of the millennium neared, Western scientists and Indigenous Peoples alike began asking a deeper question: Do bison belong among us in some larger way? Has the time come to stop treating the buffalo as a relic of a lost past? Should we instead consider plains bison restoration as both a key dimension of Indigenous reconciliation and a vibrant part of our shared ecological future? If so, we should allow them to perform their ecological function, for bison are a keystone species that shape the landscape to the benefit of others through their grazing, wallowing, defecating, and hair shedding. We should also treat them in the wild like other wildlife, allowing them to run free. Buffalo should once again be a physical presence in the landscape, in recognition of their central role in the lives, economy, and culture of Indigenous Peoples. Almost simultaneously, the idea of restoring the ecological bison gained support in scientific circles and a number of First Nations entered into a Buffalo Treaty among themselves to advocate for the animals' return.

Banff National Park became the testing ground for these larger ideas. It was the perfect choice. Plains bison were present in the Banff landscape for ten thousand years. We know this from the bones found in the ancient hearths of First Peoples and from bison skulls thousands of years old that were found during building excavations in the town. The first European explorers saw bison in the Pipestone and North Saskatchewan river valleys. By the time the park was established in 1887, they had disappeared. But they never left the landscape; backcountry park wardens still found bison bones on the ground from time to time.

BANFF IS THE world's third-oldest national park and is associated with the very beginnings of global wildlife conservation. A few surviving bison were gathered in the 1870s by far-sighted people who saw that the end of the species was near. In Canada, Charles Alloway (a hide buyer) and the Honourable James McKay (Speaker of the Legislative Council of Manitoba) brought some live calves back from a Métis bison hunt in Saskatchewan. Eventually, they were sent to Banff and put on display in a fenced-in grassy area at the foot of Cascade Mountain that became known as the Buffalo Paddock. This was the first endangered species conservation project in Canada, and among the first anywhere. But that was only the beginning of the fascinating story of plains bison recovery in Banff.

A few hundred kilometres straight south, in the grasslands of the Flathead Valley of Montana, a bison herd was established thanks to the efforts of an Indigenous man named Latatitsa. Through his advocacy, the Salish-Pend d'Oreille people decided on one of their last bison hunts to bring some surviving calves from the Great Plains back west over the Rocky Mountains to their reservation. The tribal lands were unfenced and productive, so the herd grew rapidly to over six hundred animals. It became by far the greatest plains bison herd of all and eventually became the property of a Métis man named Michel Pablo. When the U.S. government broke up the reservation into smaller parcels in 1904, the herd had to go. But the American government was not interested in buying it. So Pablo approached the Canadian government representative in Montana. Would Canada like to buy the herd?

Since Banff (then known as Rocky Mountains Park) was the epicentre of Canadian conservation, the government agent contacted Howard Douglas, the park superintendent. Douglas worked skilfully to obtain the funds necessary for the federal government to buy the herd. What followed was a wild adventure. It took three years to round most of the bison up and load them into train boxcars for travel to Canada. The greatest number went to Elk Island National Park, east of Edmonton, and most of those were sent on to a newly established Buffalo National Park, near the Saskatchewan border (subsequently disbanded; that land is now Canadian Forces Base Wainwright). One carload came to the Banff Buffalo Paddock for display. The Elk Island buffalo herd flourished in a much larger but still fenced park; it eventually became the world's most important source of disease-free plains bison for re-establishing populations elsewhere. But after a failed attempt to re-introduce bison in Jasper National Park in the late 1970s, there was no move by Parks Canada to do more. So, for about a hundred years, the only buffalo in Banff were a small display herd.

IN THE MID-1990S there was a lot of debate over commercial development in Banff National Park and its impact on the park's wildlife. Many people, including me, argued it had gone too far. A national controversy ensued that attracted international attention. To address these concerns, the federal government commissioned the Banff–Bow Valley Study, which after two years of work recommended that development come to an end. It also concluded that the Buffalo Paddock, and the nearby horse corrals, airstrip, and cadet camp, should be removed because they were blocking a wildlife corridor important for other species, especially carnivores. The bison and the fence were duly removed. The study implementation panel also recommended that wild bison be reintroduced to the park. Parks Canada undertook to study this idea, which had already been percolating inside the agency, and determined it was feasible—but no action was taken.

Restoring bison to Banff would require external activism. The Eleanor Luxton Historical Foundation started the Bison Belong campaign to encourage Parks Canada to act. We joined forces with Indigenous proponents of the Buffalo Treaty and created a social movement to engage the Canadian public in the cause. The campaign took off. In the words of the federal environment minister: "The whole country has buffalo fever." We were rewarded when the Conservative government declared they would implement the project, and the subsequent Liberal government followed through. (This story is told in more detail in the book *The Last of the Buffalo: Return to the Wild*.)

The buffalo reintroduction, which was accompanied by Indigenous ceremonies and protocols, has been a wild success. It is now a major credit to Parks Canada. As I write, there are over 150 wild plains bison roaming the backcountry of Banff National Park.

AROUND THE TIME of the Banff–Bow Valley study, Karsten Heuer was a young biologist studying the blocked wildlife corridor. He came to a talk I gave about a new conservation initiative that would link the Yellowstone National Park area to the Yukon territory to allow wild animals to roam freely up and down the Rocky Mountains. We became good friends. Karsten was an adventurer with an enormous capacity for rough travel and an appetite to initiate big projects. He asked if he could walk from Yellowstone to Yukon to promote the idea, and he did just that, raising his own money to support his journey. Along the way he reconnected with the equally tough and ambitious Leanne Allison. They would marry and go on other grand journeys together, writing books, making films, and receiving many awards for their accomplishments.

Eventually, Parks Canada would hire Karsten to lead the reintroduction of bison to Banff. This book is about that successful project. It is also the account of the final adventure of a great adventurer who, in terms of his determination to understand and embrace the wild world, was more of a nineteenth-century character than one of our times.

He died too young. His memorial service was held just outside the park in Canmore, where he had spent countless days volunteering to protect an important wildlife corridor south of the town from being compromised by a housing development. It was attended by an overflowing crowd numbering in the hundreds. Before he passed, Karsten asked me to speak at the service about his activism. I was honoured and said:

> Karsten lived life as a verb: dynamic, changing, never static, thrumming. Activism is a recognition of that life force. It's a recognition that you participate in the life of your community and that you participate in the life of all our relations. That we work for what we know that we love and for what matters to us.

Karsten knew what mattered to him. How we treat bison in the future will say a lot about what matters to us. We have it in our power to right a major historical wrong by restoring them to the broader landscape, free to roam as wildlife. If we do, it will also say a lot about our commitment to reconciliation with Indigenous Peoples. Meanwhile, we can celebrate that bison have returned to Banff National Park where they belong. We can also be grateful that Karsten wrote this fascinating book to tell us how he and the Parks Canada team were able to make it happen on the ground.

HARVEY LOCKE

CONSERVATIONIST AND WRITER

AUTHOR'S NOTE

ONE OF THE things I admire most about bison is how confident and relaxed they are; unlike humans, they are at peace with who they are and what they embody. This is likely due to their physiology—why worry about anything when you're North America's largest land mammal? Big-headed and broad-shouldered, bison are designed to meet challenges head-on.

Despite this, my team and I frequently underestimated these animals while trying to reintroduce them to the backcountry of Banff National Park in the Canadian Rockies. We assumed this group of plains bison, transferred from Elk Island—a small, flat, fenced, and predator-free national park in central Alberta—would be unprepared for life in the mountains. But these magnificent animals buffaloed us. They hoodwinked and surprised us wildlife biologists again and again. They defied our expectations and taught us innumerable lessons about the herd, about the human relationship to nature and science, and about ourselves.

In the late nineteenth century, tens of millions of wild bison were obliterated from a huge range that covered half of North America. That was a century and a half ago—about ten bison generations. That's enough time for a species that almost went extinct to forget how to be wild. To reintroduce bison to a mountainous corner of their former historic range, I assumed we would have to retrain them to fend for themselves, to navigate steep terrain and flooding rivers, endure fickle weather, and face their old foes: wild grizzly bears and wolves. I was wrong. The bison tapped into something that took me a while to comprehend during the project: something extraordinary that defied science. Although I began the reintroduction thinking we had much to teach bison, by the end I realized bison had much to teach us.

I always knew I wanted to share the story of what unfolded in the backcountry over those seven years, but I lacked the time to write it down—there was simply too much work to be done. However, after a serious accident, I developed a terminal brain condition that suddenly cleared my plate. With a short period of time left, I began to write. Here are the bison lessons I learned while working to reintroduce the animals to Banff National Park.

KARSTEN HEUER
BISON REINTRODUCTION PROJECT MANAGER (2015–2024)
BANFF NATIONAL PARK, OCTOBER 2024

A NOTE ON TERMINOLOGY: I use "buffalo" and "bison" interchangeably. "Bison" was originally a Greek noun meaning "ox" and is the animal's scientific name today. "Buffalo" is derived from *boeuf*, the word early French explorers used for "ox." Some believe "buffalo" should only apply to certain African and Asian ungulates, but the term, used by North American pioneers and explorers, appears frequently in historical accounts. Most importantly, it is used widely by Indigenous people today.

PROLOGUE

ALL MY DECADES of training and experience converged in this moment. The bull bison stared back at me from where he stood, fifty metres away, amid the bushes of the shrubby meadow where I had surprised him. My horse danced anxiously beneath me. Tapping into twenty years of backcountry warden experience, I nudged my horse back into the forest, behind a screen of evergreen trees. Dismounting, I tied him to a trunk, and, watching the bull through the branches, I reached into the scabbard and pulled out my gun.

The bull just stood there, neither fleeing nor approaching, leaning curiously toward me, his black eyes glinting beneath a mop of curly dark-brown hair. He looked healthy but forlorn, as if he were asking, "Are you my friend?"

"Not today," I muttered under my breath as I fumbled three bullets into the magazine and inhaled deeply, trying to settle my racing pulse. This was going to be tougher than I thought.

I had ridden up the Clearwater Valley to that meadow because of the reports we'd received about a bison surprising backcountry equestrians at their camps. He must

have been on a wander from the rest of the herd of plains bison we had brought into nearby Banff National Park three years earlier. He was doing what young males do: searching new areas for possible mates. In so doing, he had inadvertently stepped across an invisible political boundary, out of the park onto public lands managed by the government of Alberta, where bison are unwelcome.

"If Parks Canada reintroduces wild plains bison to Banff," the Alberta government had said when the project was under consideration, "it needs to keep the animals to the park. You need to deal with any that get out. We're already stretched."

Anyone who knows how rugged the mountains are here can appreciate the challenges inherent in such a constraint,

never mind trying to control the urge that drives bison to explore. Despite these seemingly impossible conditions, our Parks Canada team forged ahead with the project, employing some unconventional methods to keep the bison where they were supposed to be, like drift fences and herding the animals through the mountains. These strategies worked, but still, the occasional bison leaked out of the park.

"Not a great ambassador for wild bison," I said to colleagues as complaints about the bull trickled in. One complaint came in the form of a video posted on Facebook, which showed the bison approaching tethered horses as they fed on hay. After almost touching noses with them, he taunted them to play. The bison did no harm, but his behaviour was concerning.

For the next two days, we looked for him from the helicopter. We saw him once, but he disappeared before we could land. The best way to catch up with him, I figured, would be from the ground. I arranged for a truck and trailer with two horses to come from the Parks Canada barn in Banff the next day.

I had a lot of time to think during the four-hour drive north to the trailhead, and the entire day of riding in. The area is part of the remote Alberta backcountry along the Clearwater River, just outside the rarely visited northeast corner of Banff National Park. On my way, I drove through the rural towns of Sundre and Caroline, where just about everyone passing me in their pickup trucks probably would have traded anything to be the first person to hunt a wild bison in Alberta's Rocky Mountains in over a century. Instead, I was filled with dread.

WE CALLED IT the "elephant gun" during our training sessions and practised so we could use it to put aggressive grizzly bears down. The recoil packs a huge punch and always leaves a grapefruit-sized bruise on my shoulder. Not so with the other rifles or shotguns we used to finish off black bears or elk that had been hit along the Trans-Canada Highway or on the railway in the busy Bow River Valley. As millions of people every year visit that part of the park and the town of Banff for skiing, hiking, and nature tourism, we purposefully reintroduced bison to a more remote section of the park: the lesser-known headwaters region of the Dormer, Panther, and Red Deer rivers, just south of the Clearwater. The mountains—though less rugged and without the glaciers or mightier peaks of the tourist region—are wilder there. Comprising 1,200 square kilometres, the area where we hoped to anchor the bison sees fewer than a hundred people per year.

The weather was hot and the flies were bad when I arrived at the Alberta Forestry cabin where I had arranged to spend my first night as I looked for the bull. As I opened the door, I was hit with a stale, mousy smell that brought back memories of the last time I stayed there, when I had to clean rodent droppings from the cupboards and was kept awake by scurrying all night. Fortunately, someone had cleaned recently, and I had a restful sleep. I was up, packed, and on my way early the next morning, searching the old forests, swamps, and meadows as the sun cast a soft pink light on the surrounding peaks.

I looked for the young bull in all the likely places, visiting two tent camps of horse riders along the way. Neither party had seen the bison, so with the time approaching noon, I decided to check the last place we'd seen him from the helicopter. And that was where I found him, looking listless and very much alone.

HE CONTINUED TO watch me as I walked back into the meadow with the loaded gun. I found a clear line of sight, and as two ravens flapped their wings overhead, I sat in the grass and assumed my favourite shooting position: elbows braced against bent knees. I lined up the crosshairs of the rifle's scope with the bison's deep chest, muscle memory from all the training taking over. As I waited for him to turn, I breathed deeply, calming my galloping heart.

"Squeeze; don't slap the trigger," I reminded myself. This was no time for a "flier," a missed shot due to flinching in anticipation of the gun's recoil. I visualized my target: a bald spot in the shaggy hide near the bottom of his chest, rubbed smooth by the animal's elbow. Behind that mark was his beating heart, as big as a cat.

It was while I studied him through the rifle scope that I realized which animal he was. The small yellow tag embedded in his left ear gave it away: Number 23. The calf I saw born in the holding pasture. The tag was broken, a stub that could only be spotted through the rifle scope, and seeing it condensed all the moments I had witnessed of his life: his birth on a snowdrift; the many chases with other calves in the holding pasture; the comfortable way he and his herd mates stood on a ridgetop the day after their release. I thought of how, over the following months, the ancient wallows and trails the herd reactivated also unearthed their ancestors' bones.

His stare finally broke, his nose twitched, he took a step to the side, and the bald spot came into view. With a deep inhalation, I wrapped my finger around the trigger and squeezed.

LESSONS FROM THE REINTRODUCED BANFF HERD

1

LEARN FROM FAILURE

LONG BEFORE THAT wandering bull's life began, I had to figure out how to reintroduce plains bison to Banff National Park, with the aim of returning the animal to its important ecological and cultural roles. I knew very little about the species. With no blueprint to follow, my best guide was an example of what *not* to do, courtesy of a failed attempt in nearby Jasper National Park forty years earlier. The animals had bolted days after being released, travelling four hundred kilometres until they finally stopped in a farmer's field just outside the town of Grande Prairie in northern Alberta. It was a shame, but not a surprise—globally, more than half of all wildlife re-introductions fail.

Our goal was to not repeat the mistakes made in Jasper. The projects were similar in several ways. Our bison would also be sourced from Elk Island, a small national park on the eastern and flatter side of Alberta. We planned to truck them about four hundred kilometres and then, like in Jasper, airlift them the final twenty-five kilometres by helicopter to a backcountry pasture. At that point, though, our plans diverged. First, my team—myself and three others—aimed to hold our animals in that home pasture for eighteen months, not just forty-three days, as they'd done in Jasper. Second, we planned to start with fewer animals—sixteen instead of twenty-eight—and focus on young, pregnant females, not bison of all ages, so they would calve soon after arriving and anchor to their new surroundings. Third, unlike in Jasper, we intended to ceremonially prepare the animals for different stages of the reintroduction with help from Indigenous Nations, who revere bison for their historical role in sustaining their ancestors, both physically and culturally. Finally, to address the Alberta government's unwillingness to accommodate bison outside the park, we planned to discourage the animals from leaving by constructing drift fences: finite sections of wire strategically placed between natural barriers such as cliffs and heavy timber.

The drift fences were controversial. Some people thought it was a heavy-handed approach for a national park that is mainly unfenced, especially in the backcountry. The backcountry is rugged, remote, and without development, apart from some flood-damaged trails and old park warden cabins. It is a place where wildlife outnumbers people, and those who appreciate its grandeur want to keep it that way—a place where everything moves freely. This matters to me too. I have devoted my life to wildlife connectivity, going so far as hiking 3,400 kilometres from Yellowstone National Park to the Yukon to promote the Y2Y vision of wildlife corridors—keeping the vast area connected to ensure large animals like grizzly bears will survive.

I was conflicted about the use of fences to return wild bison to their historic ranges. For this concept to work, though, fences seemed necessary, especially if we wanted to be good neighbours and honour our no-bison-outside-the-park commitment to the government of Alberta. Alberta's stance is unique among Canadian provinces. British Columbia, for example, which bounds the park to the west, recognizes plains bison as wildlife and manages them similarly to deer or elk, with regulated hunting. The same is true in Saskatchewan, to the east.

I was tired of discussing such issues at the end of a long week of stakeholder meetings in early 2016. The last thing I wanted to talk to anyone about was fences, especially on my weekend. But I couldn't escape it. At a friend's retirement party, another retired colleague, Dave Norcross, beelined over to bend my ear.

I always made time for Dave. Our careers had overlapped in Banff's backcountry, where we worked hard and shared a love for exploring untrailed areas. He had started in the Warden Service about twenty years before me and had worked on the failed Jasper bison project.

"I heard about them fences you guys are thinking of building," he began in his drawl. I mustered a fake smile, bracing myself for the usual negative response to follow, but Dave surprised me.

"Best thing I've ever heard," he said, leaning in close to slap me on the back. "I sure wish we'd done something like that."

It took me a few seconds to digest his enthusiasm.

"Really?" I asked, stepping back, a little stunned. He flashed a two-fingered salute.

"Swear to God," he said. "I followed those brown buggers for days after we released them." He recalled how they led him and his horses over several mountain passes. "They got so far that I had to rent a plane." He paused and took a sip from his drink. "A Cessna, I think. Took us a while to attach the antennas so we could track the few radio collars we put on them in captivity." He shook his head, drifting off as he reminisced. "All that country," he sighed. "All those rivers..."

Then he snapped out of his reverie, and a look of indignation swept over his face. "You know where those bastards finally stopped?"

I shrugged as Dave threw back his head and downed the rest of his drink.

"At the first goddamn fence they saw!" He spat as he slammed down his empty glass. "And it wasn't anything special," he said, shaking his head. "It was one of them diddly farmer's fences." His hand sliced through the air and bounced off his hip. "Three wires. Only came up to here!"

Dave had just given me my first buffalo lesson, and I took it to heart.

2

SEIZE
OPPORTUNITY

BISON CAN'T BE reintroduced just anywhere. Several stars had to align for Banff National Park to be the right place at the right time. If the project were to succeed, the reintroduced herd would be only the fifth wild population to wander where plains bison roamed prior to being hunted to the brink of extinction.

The first thing you need in order to reintroduce wild bison is a large wild area. That is a limited commodity in today's scarred natural world. Human developments have consumed more than 75 percent of the planet, and wide-ranging wildlife are being squeezed out. On a Google Earth image of the ecoregion of the Great Plains, for example, almost all the historic range of plains bison—that area between the Rocky Mountains and the Great Lakes and south to Mexico—has been converted into a patchwork quilt of agriculture. In Canada, 80 percent of native grasslands are gone.

Much of Banff National Park, tucked in a mountainous corner of this tragic narrative, has escaped such development. The landscape is too steep and rocky for the usual roads or towns to be built there, the climate is cold, and animal habitat is legally protected. On the flip side, it suffers from overtourism—more than four million people annually come to see its spectacular scenery, mostly in the Bow River Valley. Fortunately, this accounts for a small portion of the park, about 5 percent. The remaining 95 percent represents a rare opportunity: a chance to bring back an echo of one the greatest wildlife sagas on Earth.

It would be wrong to pretend that returning bison to Banff could ever reverse the damage wrought by the overhunting that led tens of millions of animals to the brink of extinction in the nineteenth century. No one, including the government of Canada, is trying to turn back the clock on the wave of development that now blankets much of the historic range of bison with roads, towns, and farms. But a much smaller initiative—returning a few hundred of the shaggy beasts to an already protected area—could be feasible. In doing so, we might revive some of the bison's ecological and cultural connections that would otherwise be lost for good.

Banff provided an area large enough to try such an initiative. The park is 6,500 square kilometres, or about the size of the Greater Toronto Area. Quadruple that and we have the size of the Canadian Rocky Mountain Parks World Heritage Site, of which Banff is a part. Multiply Banff Park by two hundred and we step into the realm of the Yellowstone to Yukon (Y2Y) ecoregion, a network of protected areas and wildlife corridors that includes Banff. Y2Y is known as one of the most intact mountain ecosystems in the world, where healthy populations of bears, wolves, and every other large native mammal, except plains bison, still roam.

So we had access to a big area, connected to other sizable natural areas. But that was only the first step in finding an appropriate place to reintroduce bison. The chosen area also needed to provide good habitat. An assessment of Banff proved it to be worthy. The meadows that stretch into the mountains from the foothills could support hundreds of bison, maybe even a thousand. A population that size would avoid the long-term risks of inbreeding that sometimes plague smaller groups, or the risk of dying out from extreme weather events or diseases.

Political and legal support for bison is another pre-requisite for any reintroduction. Inside the national park, Banff enjoys this in spades. Canada's National Parks Act clearly states that the primary goal of national parks is to maintain and restore ecological integrity. This includes reintroducing species that, like bison, are now absent but belong here (as evidenced by many archaeological sites throughout the park where bison bones up to thousands of years old have been found). Reintroducing bison also restores processes like grazing and nutrient cycling, with positive effects that ripple through the ecosystem. These ecological relationships include interactions with Indigenous Peoples, whose rejuvenated practices, like burning and harvesting, help maintain an area's biological health.

The political and legal support for wild bison had yet to extend into Alberta. With the reintroduction project, Banff would enable the first free plains bison to roam since the province was established in 1905. That was not long after the big wild herds, which were known to delay passenger trains for hours, were wiped out by Europeans overhunting them for their meat, hides, and for so-called sport. The population of Indigenous Nations who depended on the animals also dwindled. Like the last few buffalo that were rounded up by European settlers, Indigenous people were relegated to reserves. Their fate, which had always been intertwined with bison, continued to parallel that of the animals in this dark chapter of Canada's story. The tools of control and repression included fences for bison and residential schools for Indigenous children.

Today, the government of Canada has committed to the process of truth and reconciliation, which seeks to renew the relationship with Indigenous Peoples and commits to halt and remediate damage from such colonial atrocities. This remediation includes reintroducing buffalo and its many lost links to the land and Indigenous culture. A good example is the modern Buffalo Treaty, an Indigenous-led agreement that was established in 2014 and already has more than fifty Indigenous Nations as signatories. Together, they are committed to bringing bison back to their traditional homelands. Some nations have captive herds, like those kept on the Kainai, Stoney, and Assiniboine reserves of southern Alberta and northern Montana, for food security and traditional harvest. Others are focused on reintroducing wild bison to roam sacred sites, such as Chief Mountain in Montana. Some Indigenous Nations, including the Blackfeet in northern Montana, invite co-operative efforts between tribes, conservation groups like the World Wildlife Fund, and government bodies like the U.S. Department of the Interior to ensure bison will once again roam on Indigenous, public, and national park lands.

For a bison reintroduction, there must also be a source of healthy animals to draw from. Fortunately for Parks Canada, we already managed such a herd in Elk Island National Park, about four hundred kilometres northeast of Banff. Surplus animals from that herd, rounded up every two years, have been used in conservation projects for decades. Numbering around four hundred animals today, the herd descends from eighty-one individuals that were captured from three of the last wild herds to occupy the plains in the late 1800s. In 1907, after that herd had grown to several hundred animals again, it was bought by the government of Canada, and sixty of the animals were transferred to Elk Island. They have been managed carefully by Parks Canada ever since then and are now the closest things to living museum specimens of the last wild buffalo that exist.

The elements for a successful reintroduction of bison to Banff had aligned like bullets in a chamber: a big enough space, supportive policy, the resurgence of buffalo culture, and a source of healthy animals descended from their wild ancestors. The only thing missing was a trigger, and someone to pull it.

≡ 3 ≡

LOOK FOR
TRIGGERS

EVERY IDEA NEEDS a catalyst to become a reality: something to set it in motion. Banff's trigger for wild bison came in 1997, when the hundred-year-old Buffalo Paddock, where a small number of captive bison were held for viewing by tourists, was ordered closed. Research had identified the enclosure as a blockage to an important wildlife corridor around the nearby townsite, which was especially significant for wolves. The paddock's 2.5-metre-high perimeter fence prevented wolves from accessing elk, their major food source. As a result, the elk became overpopulated and began to overgraze plants that other animals, like beavers, also relied on. This set off a cascade of negative effects among other animal populations.

The impending paddock closure prompted many suggestions about what to do with the captive bison.

"Let them go!" some said, citing archaeological evidence and references to the journals of explorers to support the

claim that wild bison belonged in the area. Known to move slowly, Parks Canada committed to do a feasibility study about a wild bison reintroduction. Meanwhile, the captive bison were quietly trucked away.

The story might have ended there were it not for a small not-for-profit group—Bison Belong—which applied pressure on Parks Canada to follow through. The feasibility study showed there was enough habitat for a bison population large enough to fulfill vital ecological and cultural roles. The government of Alberta's no-bison policy was a problem, but at least Parks Canada could get started.

When the funding proposal for reintroducing wild bison to Banff was first submitted to Parks Canada's national office and vetted against conservation projects at other national parks, it scored poorly for "sustainability" because of Alberta's no-bison stance. It might have died

there, near the bottom of the pile, if politics hadn't intervened. It was 2011. The federal Conservative Party and their leader, then–Prime Minister Stephen Harper, needed a good-news environmental announcement.

Bison Belong had kept the project on the political radar. An initiative of the Eleanor Luxton Historical Foundation, the group had recently flooded parliamentary offices with thousands of postcards for one of its pro-bison campaigns. Magically, the reintroduction proposal moved from the bottom of the pile to the top. A few weeks later, in January 2012, federal environment minister Peter Kent came to Banff to announce $6.5 million in funding to reintroduce bison.

The trigger was pulled.

4

ACCEPT FATE

I WAS THE BIOLOGIST who completed the wildlife corridor study for Parks Canada in the 1990s that recommended the Banff Buffalo Paddock be closed. Although I didn't know it then, I indirectly started the process that initiated the wild bison reintroduction I would go on to lead twenty years later, when it finally came time to get hooves on the ground.

"It's as if the bison chose you," an Indigenous friend said when he heard the convoluted story years later. Considering it all, I see what he meant.

Fresh out of university, I had landed the wildlife corridor contract partly because of how little money I was willing to charge. I hired a few assistants, and we followed my simple methodology: Hike between potential barriers at obvious squeeze points—between houses, highways, fences, and cliffs, for example—and map animal tracks that crossed these squeeze points by following their footprints in the snow. The pictures that developed after a few winters of this work were compelling, especially our discovery of the route around the Buffalo Paddock and how its fence blocked wolf movements.

That study changed my life in other ways. Parks Canada's adoption of my recommendations was significant to me as a newly minted biologist, but seeing how quickly and extensively wolves responded to the removal of the paddock made an even more lasting impression. Within one year, wolf use in that area increased by 700 percent. I realized that a single individual could make a difference, and this became a rallying call for the rest of my life.

I oscillated between science and activism for the next decade and a half, maintaining a job as a seasonal park warden in the summers to help pay the bills. It was fulfilling work, but I kept looking out for other ways to contribute to wildlife conservation. When I heard about the Yellowstone to Yukon initiative a few years later, I was enthralled. Y2Y's vision to connect and protect reserves along the Rocky Mountains with wildlife corridors was so meaningful to me that I decided to hike its 3,400-kilometre distance, as a wolf or grizzly bear might, to see if the concept was possible. I discovered it is. I shared that finding in hundreds of presentations and media interviews in the many communities along the way. A longtime friend, Leanne Allison, joined me for the second half of that journey. In her I found someone willing to embrace the wild as I did, and we married soon after completing the hike. After our wedding, we moved to northern Yukon, where I worked for a few years in Ivvavik National Park, which is known as "the place where life begins" in the Inuvialuktun language of the region. The park protects the Canadian

portion of the 150,000-member Porcupine caribou herd, and it was where Leanne and I first felt the energy and power of a herd of wild animals for the first time.

Most of the Porcupine caribou herd's sensitive calving areas sit slightly past the Alaska border within what is deceptively called the Arctic National Wildlife Refuge. The six-thousand-square-kilometre Coastal Plain, wedged between the Brooks mountain range and the coast of the Arctic Ocean, was not included in the greater refuge when the land was protected, because of what lies beneath: the largest known reserve of oil in the United States.

The push to drill for oil under the calving grounds had been explored in dozens of documentaries and articles in terms of the billions of barrels of oil to be gained; the hundreds of jobs to be created; the millions of dollars to be made. Until then, the caribou's point of view had not been addressed.

Leanne and I decided we could better understand what was at stake for the animals if we migrated, on foot, with the herd, from their wintering area to their calving grounds and back again. A joint migration was a way to document the value of this place from the caribou's perspective. We would endure what the animals endured: travelling thousands of kilometres, navigating hundreds of snowdrifts, crossing dozens of rivers, and managing wolf packs, grizzly bears, storms, and dense clouds of bugs to reach the calving grounds and then return. To top it off, Leanne filmed the caribou herd while we were being herded ourselves—finding focus, exposure, light, and heart as we forded streams and walked for days on end to keep up with the herd.

That journey was a deep five-month-long dive into the herd's life that forever changed us. It uncovered an

innate animal-like potential in each of us that had become buried beneath the distractions that crowd modern life. Immersed in the herd, we felt like the unused ends of our senses regrew, widening our view of the world again, which had imperceptibly narrowed over the years. For example, a barely audible thrumming sound emitted by distant animals, which we felt as much as we heard, came to guide us whenever we lost the herd. They also increasingly informed our route-finding decisions through hints we received in our dreams.

This herd wisdom, or group consciousness, enveloped us like an invisible cloud. It was bewildering to me, especially given my training as a scientist. As a result, I had difficulty with the transition back when it came time to return to regular life. I did my best to honour the experience by writing a book, *Being Caribou*. Leanne produced an award-winning film of the same name. But everything special that had happened was soon lost in the crush of everyday life. We carried on like many people in their thirties—we committed to a mortgage and had a child. I returned to my job with Parks Canada, but with each passing year, my restlessness grew. The caribou had taught us so much, and I had turned my back on it.

In a move to rectify this, I decided to tap into the patience and fortitude I had learned from the caribou to influence policies, laws, and plans that would shape the future of wildlife conservation. I left my twenty-five-year park position to lead a non-profit conservation group. I imagined myself donning a suit and tie to meet with environment ministers and heads of influential corporations for this cause.

The reality was starkly different from my vision. My new role quickly became a treadmill of the obligatory administrative tasks required to run an organization. When I occasionally broke free from the office—in said suit and tie—it was usually to raise some of the two million dollars in annual funds needed to maintain thirteen employees. The stresses, travel, and challenging personalities affected me, and, after a year and a half, my sleep and mental health were suffering. Concerned, I visited my doctor, who handed me a note for a six-week leave. I never returned.

Initially, I thought those two years had been a waste of time. But when the opportunity to lead the bison re-introduction in Banff National Park presented itself a few months later, I saw the value of the experience. The new position called for a project manager more than a bison biologist, someone to oversee a small team of employees and contractors, manage a budget of millions of dollars, and break down complex goals into schedules and doable tasks. The job I'd left because it was killing me had given me the administrative and leadership skills that qualified me for this now. My background as a wildlife biologist who also happened to know the bison reintroduction area from years of being a backcountry warden, plus my horse and wilderness travel skills, was a bonus.

I remained quiet in the job interview about my earlier role as a trigger for shutting down the Banff Buffalo Paddock in the 1990s, which had started the whole conversation about a possible bison reintroduction. I also didn't mention my experiences with herd consciousness with the caribou, or how I wondered if such group wisdom could be accessed through bison as well.

"Keep cool," I reminded myself at the end of what felt like a solid drilling. "Steer clear of the extraordinary stuff."

Little did I know how much the extraordinary would factor into what was to come.

5

LISTEN TO
YOUR ELDERS

THE FIRST ORDER of business in my new job was to recruit help. I invited two Parks Canada colleagues: Saundi Stevens, a quiet, unflappable veteran whom I could count on when things grew intense, and Dillon Watt, an honest, young, clean-cut co-worker who worked so hard his bosses struggled to keep up. Both of them had technical skills, landscape familiarity, and an interest in bringing bison back, but, like me, lacked any experience working with the animals we were to reintroduce. We needed someone who knew about bison. After searching more widely, I hired Adam Zier-Vogel from Wood Buffalo National Park. Not only did he have the experience we needed, he was organized and could help with the complex logistics of the project. As a bonus, he was an avid mountaineer, someone who could move well through the terrain. Pete White, who grew up in the Bow Valley and had a master's degree in wildlife biology, often helped us and seemed to possess the practical skills to solve any problem. Finally, I phoned Wes Olson, a former park warden who had facilitated the bison program at Elk Island for twenty-four years and later helped reintroduce bison to Grasslands National Park in Saskatchewan, where he still resided.

"I need your advice," I said that fall of 2015 after introducing myself and briefly describing the project over the phone. "It'd be great to look at the area together." I dangled the offer of a backcountry trip. "It'll take us four days on horseback."

Wes looked comfortable in the saddle when we met at the trailhead nearest to the Banff bison zone two weeks later—a spot at the end of a gravel road three hours north of Banff. Tall, blue-eyed, mustachioed, and wearing jeans and a black cowboy hat, Wes seemed like he had just stepped off the set of a Western. I was struck by his friendly and trustworthy appearance. We fell into easy conversation, and he pointed out old bison wallows in the meadows of the thatched grass where we started—an area called Ya Ha Tinda, or "prairie-in-the-mountains," by one of the local Indigenous Nations, the Stoney Nakoda. Shaped by the warm chinook winds that blow there much of the winter, and in the rain shadow of the mountains, the area is too dry for the trees that dominate the surrounding mountain valleys. It is also rich in archaeological sites, including teepee rings, ancient campsites, and at least one buffalo jump.

The more we chatted, the more obvious it became that I was in the company of one of Canada's pre-eminent bison experts. I had pulled Wes away from his work authoring plans to bring the animals back to historical parks, private conservation lands, and Indigenous Nations throughout the country. His knowledge came not only from his Parks Canada experience but from a stint as a bison rancher and a lifelong interest in all things bison. Wes has read every research report and scientific paper on bison and regularly attends bison conferences. He has synthesized this information to write several bison books and is the go-to person for bison advice among several governments and Indigenous Nations. Many refer to him as Grandfather Buffalo.

Wes was impressed with Ya Ha Tinda. Although the ancient bison had disappeared long ago, the open hillsides and grasslands make it some of the best winter range in Alberta for their still-extant ungulate cousins—bighorn sheep and elk. Some of the largest specimens of sheep and elk in the province have been killed by trophy hunters here. The buffalo wallows Wes pointed at were part of the greatest concentration of such living artifacts discovered in the Canadian Rockies.

But when we entered the park two hours later, as the trees closed in and the mountains reared up, Wes's voice went quiet. He remained silent for much of the ride that day. Other than a recently burned forest and the grass-covered flood plain of a side creek, there was not a lot of obvious bison habitat to remark upon as we rode west.

"They might like this," he mumbled when we emerged from the shadows and splashed across the creek. "Or this," he said half-heartedly when we arrived at the meadow in front of the patrol cabin where we spent the first night. Looking around that evening at what we could see of the park and the proposed bison zone, we spotted a few meadows, but it was mostly a blanket of dark evergreen trees.

The next morning, we continued to follow a tunnel cut and cleared through the trees by the trail crew that travelled it annually. It led up and over a low pass that separated the Red Deer and Panther valleys, bringing us to the next patrol cabin, in the centre of the proposed bison zone. It was here that we intended to hold the bison in a pasture for the first eighteen months, so they could calve twice and anchor to this place.

"We'll extend the posts with two-by-fours," I enthused, showing Wes the old horse pasture we aimed to convert for bison. "We'll staple up some proper game fence." He looked dubiously at the sagging wire. "You know, that metal mesh stuff."

For the next hour I toured Wes around the old warden station, trying to invite him to see my vision. It had been built in the 1950s, back when the collapsed fire road out front was still driveable to Banff. The cabin was more of a small house, with an old concrete basement for the cold pantry and a huge wood furnace. There were two bedrooms on the ground floor, a practical kitchen, a workshop, and an old tack shed and corral down the hill where we could store feed.

"There's even an old bomb shelter in case things get really bad," I joked, indicating a passageway to an underground room with worn, rusty bunk beds. In the period following World War II, every government building had to have a such a shelter, regardless of location. Wes forced out a small laugh but remained serious as we climbed the stairs and spilled out the back door.

"I can see how it might work," he said, still scowling at the extensive shrubbery that was meant to be the pasture, "but it's going to take an awful lot of feed."

"What about extending the pasture there?" I pointed to a grassy meadow at the base of a rocky avalanche path. Each winter, snow slides cleared the shrubs and any new trees encroaching into the area. Wes shook his head.

"Avalanches might take the fence out," he warned. "Then you'd really have problems."

I nodded and plied him with more questions, thinking about what the bison might do once we released them.

"Do you think they'd go there?" I gestured to a jumble of half-fallen silvery trees, a prescribed burn from twenty-five years before. Enough time had elapsed for many of the tree roots to rot, and half the forest was on the ground. It looked like a giant game of pick-up sticks. Wes grimaced.

"They won't clamber through that," he said.

"How about there?" I nodded toward a little canyon. A pocket of green grass lurked in its moist folds. Wes did not hold back.

"Not a chance," he said. "Too steep! But maybe there." He pointed to a bench of grass and shrubs that the creek formed just before spilling into the river. He searched for more hopeful ground, but all that stared back was a tilted canvas of angled lines. The evening sun brushed across the complex terrain, painting every gully in black shadow.

"This isn't an ideal place for bison," Wes said finally, "but it sure is beautiful."

WES GREW MORE POSITIVE about the area's potential to support bison the following morning. After an hour of riding downstream, we emerged from a shadowy constriction of rock and old-growth timber onto grassy bowls and benches shining in the warm sun. Littered with old antlers, it was where many elk had wintered back when their population was high, before wolves returned to the park in the late 1980s.

"Wouldn't you know it," Wes joked, as ripened seed heads from the grasses brushed our stirrups and caked our boots, "we finally see good bison habitat and it's covered with wolf scat."

For the next hour we rode side by side, speculating on what signs of wolves meant for reintroducing bison.

"Bison won't just run away like deer or elk," Wes said. "They're formidable animals. They often stand their ground."

THE MOUNTAINS CLOSED around us again as we entered yet another tunnel of trees hemmed in by towering rock walls. It took several hours of riding through the shadows to reach our third patrol cabin, where we had a final rest before continuing out of the park the next day. In the morning, the horses had to wade several times across the narrowed river. A shadowy canyon finally spat us into another grassy, sparsely treed opening. It felt like we were travelling through a birth canal, and I relished the sun on my back when we emerged from its confines. Waiting for Wes to catch up, I glassed the open slopes for elk and sheep.

The sound of his approach was a sign of the changes underfoot. Instead of the clatter of metal horseshoes on rock, his horse's hooves drummed a beat deep in the thickening soil. We had re-entered the Ya Ha Tinda area, where the grass grew in dense, well-grounded mats, not the sparse clumps that covered the few openings we'd passed through in the park.

"This is more like it," said Wes when we stopped for lunch. We could see how much better the turf was by how the horses grazed. Instead of nibbling at the scarce grass in the park, they tore this away in chunks, wrapping their tongues around wide blades and throwing their heads back to rip it off in big mouthfuls. "Bison are large," Wes said as we watched the horses chew and swallow. "They'll need quantity as much as quality. They'll be drawn to a place like this.

"That's a problem," he continued, after taking a sip from his water bottle. "I've thought a lot about what you're tasked to do here; it's complicated." He looked around in all directions at the surrounding mountains. "This landscape is complex. It's not like Grasslands Park. Bison habitat here is patchy. There's the occasional grassy meadow with swaths of not-so-good habitat between them. So much forest and rock!" He spat out the words, then gestured at the grazing horses. "The good stuff is here, just most of it is outside the park."

I nodded. He was right. It was not ideal.

"The key is to think of your bison zone in the park like a rancher thinks of their spread," Wes said as we tucked into our sandwiches. I waited for him to continue after he swallowed his first bite. He turned to fix me with his piercing eyes. "Treat it like a *system* of pastures."

"Pardon?" I said, unsure of what he was suggesting.

"Like a target," he clarified. "Your holding pasture is the bull's eye, like on a dartboard. When you release the animals from there, *you* direct them. You limit their explorations to one circle at a time away from the bull's eye; first you contain them to the nine-point ring."

I nodded.

"Then, a few months later, you let them explore the eight-point ring," he continued, "and so on. Gradually, they get to know each new pasture and become attached to it." He pulled out a piece of grass and then picked his teeth with it. "Just like that," he smiled.

I struggled to imagine implementing this across a vast and porous landscape, with such headstrong animals.

"I think I know what you mean," I said, unsure, as he continued to chew the grass.

"Have you ever played poker?" he asked.

I nodded.

"Stakes are high and you've been dealt a bad hand." He looked at me over the rim of his sunglasses. "I've seen the place these last few days," he said, "you've been dealt a bad hand indeed. So what are you going to do?"

I shrugged.

"You're going to need the whole playbook. Fences, herding…" his voice trailed off. "You name it. You might even have to trick the animals and use some bluffs. You'll somehow have to convince the animals to stay in the park until they adopt it as their own." He laughed, nodding at

the grass that surrounded us, extending all the way to where our truck and trailer was parked, a few hours' ride away. Wes reached down and grabbed a handful of grass, shaking the dirt from the roots of the thick clump. "Somehow you'll have to prevent them from leaving the park for this!"

= 6 =

BLUFF WHEN NECESSARY

THERE IS A JOKE among bison ranchers: The only kind of fence that will hold back bison is one that keeps them where they already want to be. According to Wes, the park was not where bison would want to be, and we would have to convince them it was good enough. For Wes's plan to work—gradually directing the animals out from the holding pasture to other meadows, as if they were other pastures—we needed some tools to help separate those meadows and steer the animals between them. Fences were key, but being in a national park, they had to be designed and constructed so that other wild animals could travel under, over, or through the obstructions.

I pondered this tough-to-solve riddle for a long time, always keeping in mind what Dave Norcross had said about where the escaping Jasper bison had finally stopped—at a diddly farmer's fence. I also thought a lot about what Wes said about maybe having to use a bluff or trick. No amount of engineering would hold back a determined one-tonne bison. Maybe creating a few well-placed bluffs was all we could do to convince the bison to stay in the park. It helped that the animals were coming from Elk Island, where the park is surrounded by a 2.5-metre-high mesh fence. The animals would have learned to respect the fenceline, and we could renew that respect before releasing them by electrifying a similarly designed fence to retain them in the holding pasture.

We began with a fence design chosen by university and government of Alberta researchers who had worked on this same riddle years before. Their focus had been to hold privately ranched bison on leased provincial lands while still allowing deer and elk to pass through. The successful design served as a good starting point for us, something not too different from a regular farmer's fence. It measured a little taller—just above waist height on an adult human—and consisted of five wires instead of three, all smooth, with no barbs.

We built and tested a few short sections, experimented with wire heights, and monitored wildlife reactions with motion-activated cameras. After photos from wind and other false triggers were removed, the 1.5 million images were reduced to 3,700 wildlife–fence interactions that showed all species moving over, under, and through them. But we were unsatisfied, because a few animals were still being deflected. We finally opted for an adjustable design that could be switched between two modes depending on the location of the bison. If they were far away, the fence would be in wildlife-friendly mode, with the five wires pinched in two bundles, so the top wire was lower and the bottom wire higher than a typical farmer's fence. When the bison were nearby, the fence would be changed to bison-deflection mode, with all five wires deployed.

With the help of some hardworking and creative fencers from the nearby town of Sundre, we built fences in thirteen different places to influence bison movements within a 1,200-square-kilometre zone in the park. I chose spots with natural constrictions between Wes's imagined pastures, augmenting and tying into natural barriers like cliffs and thick forest, and running lengths of wire that ranged from a hundred metres to two kilometres. Fence adjustments from one mode to the other were made by hand, using fencing staples arranged like cotter pins on posts. The posts primarily consisted of live trees that were protected with scraps of untreated lumber. In the absence of trees, we used metal posts made of recycled oil-field drill stems. In places where the ground was too rocky for drill stems, we used posts of metal U-channel, which, when sledgehammered, usually penetrated the hard ground. The work was challenging, but we got it done. We spread the thirteen projects over two years before the bison were scheduled to be transferred.

Building fences across mountain rivers and creeks requires innovation. Those watercourses fluctuate wildly with rain and snowmelt. We initially tried hanging curtains of polypropylene mesh or screens over the water, but they caught too much wind and billowed to the point of pulling out posts. After lots of experimenting, we opted for tennis nets. They were durable and light, didn't catch

too much wind, and, most importantly, I found a supplier in Calgary who dealt in used ones for a reasonable price.

"You want them for what?" he asked when I tracked him down in the gritty, industrial part of the city. "I have them because of pickleball," he explained when I questioned why he had so many used nets. "Pickleball has replaced tennis as the number-one court sport. I couldn't bear to throw out the old nets when I did the conversions. Some might be a little tattered," he allowed, "but they're otherwise perfectly good. I knew someone would come along one day who could use them."

———

I REALIZED HOW MUCH our bison fences differed from what people expected when I escorted two colleagues from Wood Buffalo National Park to see our layout a few years later. The Wood Buffalo team were thinking of using a fence to discourage healthy and diseased bison herds in their area from mixing. They wanted to see one of our bison fences to get ideas for how they might approach their unique situation. I remained silent as they brainstormed in the vehicle on the way to the most accessible fence, at the end of a restricted four-wheel-drive track where the Ya Ha Tinda Ranch gives way to the park. Their fencing ideas sounded like Jurassic Park, with huge metal posts, concrete footings, and solid steel. I decided to wait until they saw our fence so the design could speak for itself.

They were speechless when we arrived at what seemed like a typical farmer's wire fence with smooth, not barbed, wires. It stood a little higher than a fence you might see alongside a country road, but not by much. Without a word, we wandered up and down the line for a while, past some tree posts where I demonstrated the simple cotter-pin system of three fencing staples we had attached to adjust the wire heights. I proceeded to walk them to the river, where the wires changed to tennis nets.

"Well," I said as I absent-mindedly tugged at a tattered section of the tennis net hanging across the river, "this is it."

"This is it?" they repeated, wide-eyed as they drank in the flimsy array of tree posts that snaked up the forested slope on the opposite bank.

"Yup." I pulled at the thin wire that held up the nets, bending it a bit with my bare hand. "I know it's not what you'd expect a bison fence to look like. It's not what I expected either." I pointed at fresh deer tracks to either side. "We had to build something that allows other animals to get through, not something that would physically hold back a determined bison. It's a compromise," I allowed, "not a sure thing. A bluff."

7

LESS IS MORE

FOR THE JASPER reintroduction in 1978, the project organizers had translocated twenty-eight animals of different ages, equally balanced between male and female. They were looking to replicate the makeup of a natural herd. The strategy backfired: The older animals failed to anchor to the new area and fled, leading the younger animals to follow.

The 2005 reintroduction of bison to Saskatchewan's Grasslands National Park took an alternate approach. They started with seventy-one very young and highly impressionable animals: thirty female and thirty male calves, plus eleven yearlings. The planners reasoned that this was enough to insulate against future inbreeding. It was like turning a bunch of teenagers loose, but in a place where trouble was hard to find. Unlike in Jasper or Banff national parks, the 181-square-kilometre reintroduction site in Grasslands had no predators, there was grass everywhere, and the area was completely fenced.

The plans I inherited by the time I started on the Banff project were similar: reintroduce fifty young animals—

twenty-five female and twenty-five male calves—and hold them in a pasture for a few months before their release. However, the more I spoke with bison ranchers, the more I became convinced that our best chance of success lay in a different strategy.

Originally, I wasn't even seeking advice on what sort of animals to start with. I'd phoned up ranchers and even flew two of them to the Panther Valley to confer on how big the holding pasture should be, where to put gates and water troughs, and what kind of fencing to use. Invariably, though, the conversation would turn to the sex and number of animals we aimed to start with and how quickly they would calve.

"I worry about my fences until any new bison cows I bring deliver their first calf," said one old-timer. "Once a young 'un is born, they settle right down."

Similar advice from other bison ranchers made me stop and think.

"Twenty-five bull calves!" a rancher barked at me over the phone one day. "Are you nuts? You'll have young bulls all over the place within a couple of years! Some are bound to leave the park!"

I mulled over their comments, then pitched a new plan to my boss, Bill Hunt, in his Banff office.

"We need to change course from flying in fifty calves," I said. "Fewer bulls and more babies sooner."

Bill looked at me. I shrugged and barrelled on.

"It'll save money. Fewer animals to fly into the backcountry. The pasture can be smaller. Less feed. And, less risk." I reminded him how one of the animals had died in the Jasper translocation when the helicopter set down too hard.

As it would for any manager, a proposal to reduce financial cost and risk caught Bill's attention. Fortunately, those weren't the only things that factored in. He had become my boss the old-fashioned way—working from the field up—and because we'd both started in the Warden Service twenty years earlier, we shared a respect for each other and for the job. Bill was a practical thinker, had ridden miles on horseback, and knew the park's backcountry. And, as a biologist, he understood science. Throughout this project, we had weekly check-ins. I knew these meetings were mostly for him to keep tabs on me and the high-profile project, but I valued them as chances to brainstorm ideas with such an intelligent person. Besides, he had a wicked sense of humour and often had me in stitches. On this occasion, he listened from behind a desk piled high with papers, raising his eyebrows.

"We bring in fewer animals," I continued. "Mostly females. Two- and three-year-olds. Young, but not too young. Just sexually mature and pregnant. Preloaded." I winked. "Two for one. When the bison calve in the holding pasture, they'll anchor there. It'll be the only place the newborns know."

"Go on," Bill encouraged me.

"We still bring young bulls for breeding," I said, "but only a few. Four or five. They get busy later that first summer." I winked again. "The cows give birth a second time just before we release them, eighteen months after we first bring them in. They anchor again." I slapped his desk. "The

newborns are slower; extra insurance that the mothers will stick around. And it'll be the peak for luscious green grass. Why would they leave?"

Bill lifted a hand in surrender. "I get it," he said, chuckling. "More with less."

"Exactly!" I smiled.

"Okay," he continued. "Run me through the numbers."

"Let's say we only bring in sixteen animals..." I began. "Twelve cows that give birth twice equals twenty-four calves, plus the sixteen originals. That's forty animals at the release." I beamed. "I know it's not quite fifty, but half of them will be born in the holding pasture," I said, emphasizing this point. "They'll know nothing but Banff Park."

"How about genetics?" Bill asked. "You know," he said, whirling his hand in impatient circles, "the whole inbreeding thing."

"I already checked with a geneticist," I said. "It's safe to assume the females get bred by different bulls in Elk Island than the ones we bring into Banff. The diversity is almost as good as starting with fifty calves." I smiled again. "Like you said, more with less."

"Brilliant," Bill said. "Let's do it."

8

SIMPLIFY
COMPLEXITY

ONCE BILL AND I decided how many animals we required, I could plan the translocation. The first step was a phone call to Elk Island National Park to confirm the number, ages, and sexes of animals we wanted.

"No problem," said the manager of the Elk Island bison program, "but you'll need to help round them up next January." The small, fenced park performs biennial roundups and removes excess animals by running them through an elaborate handling facility. The park was happy that some animals from the next roundup would be used to seed a new population in Banff.

Knowing the roundup date, I worked backward from what I considered would be the most complicated part of the translocation: airlifting the bison by helicopter into the roadless backcountry. I had to ensure that each step from the roundup fed into that moment, as it was the key that would unlock everything else. For their airlifts, Jasper had used single-animal aluminum crates, which were still in storage at Elk Island. These crates were available to us and could be lifted with a regular-sized helicopter. However, as I'd said to Bill, we wanted to reduce overall risk with fewer helicopter flights. We needed to figure out how to fly several animals at once, a puzzle Bill and I tried to solve one day in his office. We got so far off topic that we were discussing custom-welded stock trailers with wheels that would pop off before the trailers were flown in. This was just one of the many hare-brained schemes we devised.

"Hey, I've got it!" Bill finally exclaimed. "What about those mini metal shipping containers?" He held up an inverted hand like a grappling hook. "They already have handles that you can hook and lift from the top; cranes do it all the time when they load them on and off ships." He started to calculate out loud. "We'd need four or five of them. Three or four bison per crate. Strap them onto small flatbed trucks, back them up to the loading ramp at Elk Island, then drive them to the end of the gravel road." He stared out the window. "I wonder how much those crates weigh? It'd have to be a big mother helicopter to lift one full of bison."

I completed more research and made a few phone calls after our meeting, then delivered my report a week later.

"Big mother helicopter is booked," I said, smiling. "One of those Russian-made things. A Kamov." I twisted my hands, demonstrating the mechanics. "Two rotors that turn in opposite directions. That's how it generates so much lift. It can easily sling small shipping containers—plus bison."

Bill smiled, and I continued.

"The nearest one is on the West Coast. They're used for logging big trees." I frowned. "They aren't climate friendly—drinks more than nine hundred litres of fuel an hour." This was a reality I had to face in a long line of imperfect solutions to keep the project on task.

I carried out the necessary arrangements in the coming weeks. I bought five 2.5-cubic-metre metal shipping containers and had them retrofitted with hatches, so we could poke a jab stick through if needed to sedate the animals, and windows for ventilation. I ordered the drugs, arranged for five mini flatbed trucks, and hired drivers, who would pick up the animals at Elk Island. Once loaded, the bison would be driven to Ya Ha Tinda Ranch via four hundred kilometres of paved highways that eventually petered into a dead-end gravel road. I charted the most efficient route and printed off five maps, one for each driver.

Thinking everything was set, I received a surprise call from the owner of the trucking company.

"I forgot to tell you something," he said sheepishly. "This might be a dealbreaker."

"What's the problem?" I asked, alarmed.

"It's the paint job on the trucks," he explained, "the company logo. It doesn't quite portray a government vibe. There's the media to consider, in case they're taking photos."

"What's the issue?"

"The sides," he answered. "They're decorated in flames."

I laughed. "Don't worry about it. Flaming bison—I like it."

THE JANUARY 2017 roundup of plains bison at Elk Island arrived quickly. About one hundred animals would be removed from the herd to keep the total below four hundred—a comfortable capacity for the small park. In the previous weeks, a series of pens were baited with hay and successive gates were closed as the animals ventured deeper into the handling facility. The roundup consisted of gently herding the animals in the pens into ever-smaller groups via a labyrinth of alleys and gates. These eventually led to the squeeze chute, a slim metal cage with hydraulically controlled walls that "squeeze" and temporarily hold each animal so it's safe for handlers to extract blood, test for disease, check for pregnancy, extract hair samples, and—on selected animals—affix a radio collar.

It takes about twenty people for the whole operation to run smoothly. My role was to work alongside the veterinarians and pull samples of hair. Dressed in vinyl aprons and rubber gloves to their armpits, the vets aged the animals by their teeth as they entered the squeeze chute, and lifted the tail of each female, inserted an arm, and palpated her innards to see if she was pregnant. If the cow was pregnant and young, I signalled the person at the exit gates to shunt her to the pen where she would be considered for Banff.

We only needed sixteen animals, but we sorted seventy bison into the Banff pen, where they were held for two weeks to ensure there was no sickness or disease. During this period of quarantine, we sent the hair samples to the local university for genetic analysis to determine relatedness and unique alleles. The university researchers provided us with a list of sixteen ideal candidates to maximize diversity, which became our herd.

The animals were about to undergo a significant transition. They were going from being semi-captive in a small, fenced reserve in Treaty 6 territory—one of the regions governed by numbered agreements signed between the Canadian government and Indigenous Nations in the 1870s—to living a freer, wilder, more dangerous life in Treaty 7 territory. The signatory nations of both areas felt the animals and the project would benefit from acknowledging this transition ceremonially.

Taking a break from sorting animals, I helped erect teepees, partook in pipe ceremonies, and listened to prayers, speeches, and songs to bless the animals and the upcoming transfer. The bison stood patiently in a nearby pen, heads hanging low, as Elders and dignitaries in headdresses gave speeches from a stage, with a hundred or so people, chilled by a late January breeze, watching wrapped in blankets and scarves. Representatives from the Enoch and Ermineskin, O'Chiese, Samson, Sunchild, and Métis Nations from around Elk Island transferred oversight of the animals to the Siksika, Kainai, Piikani, Tsuut'ina, Îyârhe Nakoda, and Métis Nations whose territories include Banff.

I was preoccupied with last-minute details for the translocation, but I couldn't help but be gripped by the gravity of what we were about to do. It was bigger, culturally, than I had imagined. Looking across the teepees and feeling the drumbeats of Native singers in my chest, I saw I was just a Parks Canada pawn in something much larger about to play out: the prophecy of bison returning from the sky and the mountains.

THE FIVE FLAMING FLATBEDS, each with an empty container strapped on the back, arrived at the handling facility by midafternoon the next day. As the first truck backed up to the loading ramp, we ran the sixteen bison through the chute system one last time, injecting every one with a long-acting tranquilizer and slipping rubber tubing over their horns to prevent them from goring one another in the containers.

I had worried the animals might balk when they turned the final corner of the chute and saw the shipping containers at the end of the loading ramp, but they continued without hesitation, three or four at a time, right into all five metal crates. Darkness had descended by the time we

shut the doors of the last container and pulled onto the highway. It was 5:30 PM. We drove slowly on the seven-hour trip to the Ya Ha Tinda Ranch and stopped only twice: once to get gas and once to check the animals. As we peered in from a ladder leaned against the top window of each container, the bison looked up, a bit dopey, into the glare of our flashlights. I was relieved to see that the horn tubes were doing their job—each one festooned with hair from their crate mates.

The house lights were on in the distance when we finally arrived at a quiet meadow at the end of the gravel road at Ya Ha Tinda Ranch. It was two AM, and Rick Smith, the ranch manager, and his wife, Jean, were waiting for us.

"The bunk beds in the guestroom are made up," Jean said, noting our weariness. "I'll take the night shift and watch over the animals." She climbed into one of the trucks, rolled down the window, and fiddled with the AM radio dial. As we returned to our truck, released from our charges for a few hours, we heard Mozart serenade the bison under a canopy of stars.

After a short sleep, we were up at seven AM, checking the weather, the animals, and logistics to ensure all was ready at the final destination. I had arranged a small crew to receive the animals at the Panther Valley holding pasture: Dillon Watt; a veterinarian; a few strong helpers; and Saundi Stevens as team leader. The day before, they had done a trial flight with the helicopter and two co-pilots, slinging in a few loads of hay.

"It mostly went well," said Saundi on our radio call that morning. "There was some confusion about weights and lift at first," she explained in her usual wry style, "but we sorted it out. Unfortunately, they hooked one of the fences at the ranch and dragged it into the trees."

That made me nervous. "Anything we should be worried about?" I asked.

Saundi's steady voice reassured me, as it would repeatedly over the course of that very stressful day. "Everyone learned what they needed to learn," she said matter-of-factly. "We're ready to fly in live animals now."

"Okay," I thought. "It's showtime."

"Let's wait half an hour," I suggested, glancing at the mountain ridge between us. "We'll see if this fog burns off."

Thirty minutes later, Saundi's voice came over the airwaves, calm but businesslike.

"Pilots say the weather looks good. Helicopter just took off. Incoming for the first container."

We heard the machine approach before we saw it, a heavy thwacking in the air, and then, from over a distant ridge, a tiny dot grew larger as the sound grew louder. In the few minutes we had before the helicopter arrived, we

worked quickly. We affixed and double-checked every clasp as we attached the four-point harness to the top of the first container, ensuring the cable was free of twists before it was hooked to the machine's longline. Suddenly, the helicopter roared over us like a giant beast, the down-wash tearing at our clothes and threatening to rip us from our tasks. We worked methodically. One person hooked the container to the longline; another checked the animals from the upper window of the container to ensure their well-being, then pulled the ladder with a thumbs-up signal to the final person, who deployed the parachute, which, once the container lifted and spun, had enough forward momentum to fill and drag behind the whole load like an anchor, steadying it through the air. I watched the helicopter shrink as it climbed and flew over the ridgeline to the west, its precious cargo dangling beneath it like a delicate knot.

"Three bulls incoming," I radioed to Saundi.

I was relieved when her transmission crackled in return fifteen minutes later.

"Three males received," she said.

We didn't have long to redo the entire setup for the next flight. The drivers assumed the next truck in line would be the one to deposit its cargo for the second flight into the Panther Valley, with the remaining three bulls. Time was extremely limited. But then I had a sudden realization: If anything went wrong and we had to suddenly stop flying, I would be a fool for having two containers of males in the backcountry. At the last minute I demanded to switch the order of the trucks. "It has to be loaded with females," I insisted.

The helicopter returned and everybody flawlessly performed their tasks as it hovered overhead again. The maelstrom of wind from the rotors hit as strong as ever. Then the machine took off, this time carrying four pregnant females.

A long fifteen minutes passed before Saundi's calm voice sounded over the radio.

"Four cows received."

I high-fived Adam Zier-Vogel beside me and shook him by the shoulders. He looked surprised as I pulled him in for a hug. We still had three more containers to fly in, eighteen months of bison ranching ahead of us, and countless unknowns to navigate—was I celebrating a little early?

"We now have a breeding population!" I slapped him on the back and shouted through the headset protecting his ears. "Just try and get this genie back in the bottle now!"

≡ 9 ≡

EMBRACE INDIVIDUALITY

A FUNNY THING HAPPENED after we flew the animals into the backcountry. What had appeared to be an amorphous group of bison back in the pens at Elk Island became a cast of sixteen unique characters once we were working up close with them inside the holding pasture. Much like humans, every animal is a sum of its genetic and learned parts. The way those parts combine to shape an individual differs greatly from one to another. Realizing this was critical to our safety while working with the animals in the pasture. It also mattered when they got out, for the survival of the individuals and the herd. Diversity equals options, and options are good when it comes to figuring out a foreign place.

At first, we were unaware of the differences among the herd; only after each of us had spent several nine-day shifts at the pasture did they become obvious. The green numbered tags we'd punched in the animals' ears back at

Elk Island made it easy to identify individual bison, and we quickly learned which ones to be wary of. Our safety protocol was to remain three metres away from any animal. If one came closer, we were to climb the fence or seek shelter in one of the containers. This worked until we inadvertently provoked one or more animals. What would set them off was hard to predict. One moment, a pair might be feeding placidly side by side, and the next, they would be locking horns and tossing each other several metres aside—no small feat considering each young adult weighed between four and five hundred kilograms. We rapidly settled on rule number one: Give the bison as much space as possible, and never corner them.

Our protocol was mostly successful, but we still earned plenty of bison warnings: They would shake their heads or paw the ground. The position of the tail, however, said the most about an animal's agitation. Held low, it denoted a relaxed state, and we could relax too; but held horizontally or higher, watch out! And when the tail pointed straight up, that bison was jacked up and about to charge. It's very unnerving to have such a large animal come at you with its horns pointed forward and eyes rolled back. Thankfully, this was not a regular occurrence, and it always ended with the animal veering away at the last second. Once out of harm's way, whoever had been charged at usually needed to sit down for several minutes to recover.

How and what triggered these charges differed with each animal. Number 10, for example, was the easiest to agitate. She was also the most dominant, always pushing others away from food and water, and intolerant of sharing space. Numbers 8, 11, 13, and 15 were similar. It became standard practice to keep the snowmobile or quad bike nearby when we threw them bales of hay, to use it as a shield in case they charged. Number 12 was more complicated; she learned to use us as a screen, accessing more and better food because she tolerated us more than the others. Numbers 7 and 17 were the most submissive, always on the periphery of the group, the last to come when we threw the animals hay. Number 6 was a leader among her own but highly sensitive to humans. As soon as we came close, she fled.

The bulls also had unique personalities. We named Number 2 Happy and Number 4 Lucky due to their easy-going nature. Number 19 was Pretty Face, the tall-shouldered bull who bred all the cows when the first summer came. Meanwhile, Number 5, the Pensive One, often stood alone by the gate for hours, gazing into the distance, as if dreaming of freedom. Little did we know that he would act on this later.

Each day we grew more familiar with these unique characteristics as we fed the animals hay, pumped water into troughs from the river, put out salt blocks so they got their minerals, and, eventually, shovelled their manure. I had not anticipated that all the dung would become a problem until I stood with our veterinarian during an impromptu inspection a month after the translocation.

"What are you going to do with all this?" He wrinkled his nose at the hundreds of brown patties piling up, especially where the animals preferred to bed. "You can't just leave it," he said, shaking his head. "The animals will get sick. We can't expect them to sleep and feed on layers of that."

I had no choice but to add manure shovelling to the list of our team's daily chores. We soon calculated that each animal excreted five kilograms of the stuff about three times a day, so we needed a system to deal with it. We were used to chopping firewood, but now, in winter, we also needed to chop frozen dung free from the ground. "Cut the crap" took on a whole new meaning. Using the snow-mobile or quad, we pulled trailers full of the brown stuff outside the pasture. Someone soon dubbed the growing pile Manure Mountain.

"Are you serious?" asked Dillon when I explained the manure management plan at our next team meeting back at the office in Banff. He swept his hand toward all the cubicles and computer screens outside the meeting room. "We already have enough BS to deal with from the national office!"

THE MOOD IN THE PASTURE clearly shifted when the bison began to calve a few months after arriving, at the end of April 2017. It was Earth Day and Adam was on shift. A little red calf, already dry, followed Number 10 out of the bushes when she came for hay that morning. Another one appeared on spindly legs behind Number 11 the next day. When I arrived four days later to take over from him, Adam pointed to where the new mothers and their calves had formed a tiny nursery group, off to the side, away from the others. They behaved differently, watching over their

new charges with a noticeable vigilance, sharing their new responsibility in shifts. While one slept, recovering from the demands of birthing and nursing, another looked out. The calves were tireless in comparison, one moment head-butting their mothers' udders to stimulate the flow of milk, and the next wrestling and chasing each other in zigzag paths, weaving around the cows like new planets orbiting tired suns.

The pasture, and the area around it, was releasing from winter's grip. A few bushes pushed through the melting snow, trying to leaf out, but it was mostly a quagmire of muck. Preparing to trudge through it on my way to feed the animals, I stopped as they came into view. Most of them waited at the gate, as was their habit, but off to the side was a lone cow, lying on a bed of snowdrifts, gasping for air in rib-jarring breaths.

My first thought was that she had been attacked or had become entangled in the fence, but then I realized she was in labour. Mesmerized, I pulled out my binoculars and quietly sat down to watch.

My vantage point was far from ideal. Bouts of warm sun were interrupted by intense snow flurries—typical spring weather in the Rockies—and the scene in front of me drifted in and out of view in waves. Perched there, I became caked in snow, only to dry out in a hot bout of sun. The cow lay unperturbed through it all, repositioning herself between periods of pushing, until a tiny snout poked from the dark recesses of her rump. Another wave of weather passed, and by the time the sky cleared, the newborn had cleared the birth canal and lay in a heap of blood and afterbirth on the snow. The relieved mother stood up, protecting her baby from above, and licked at the soaking-wet ball of fur, nudging it to stand. The two other mothers and calves ambled over to give it a sniff as it struggled to its feet. Life in the herd had begun.

That was the only birth we saw, but by the end of May, each of the ten cows had a calf, just as we planned, and had joined the nursery band. Calving undeniably strengthened the bond they had with each other, as aunties and babysitters and fellow mothers, but only time would tell if it anchored them more to the land.

That bond weakened over the following months. The qualities that set each of the bison mothers apart reasserted themselves, and as the calves grew, each one took on traits of its mother. The daughter of Number 10 became the bossiest, most dominant calf. Number 6's was the most sensitive, and the offspring of Numbers 7 and 17 became social outcasts, always lingering on the outside margins of the herd. All of them grew, nursing less and nibbling at hay and grass more, their fluffy red coats going wiry and brown.

A year later, with the approach of the second spring, came the promise of more calves. Number 19, Pretty Face, a bull that was shy with us, was obviously forward with the cows. After a lot of harmless posturing and tussles with the other males, he had emerged as the dominant bull and bred all the cows in the first summer and fall. We sometimes saw him in action while we fed the animals or shovelled their manure, but the speed at which he mated made it easy to miss. Days of following and sniffing a female as she prepared to ovulate would suddenly resolve with a seconds-long mount.

The gestation period for bison is similar to that of humans—about nine months. We continued to support the herd as the cows went through their second pregnancies, feeding and watering the born and the unborn. The second crop of calves began to arrive in May 2018. Eventually there would be ten in all, as we'd hoped, but due to the inexperience of Number 19 and the time he took between mountings, their births would be spread from late spring to late fall. This was not ideal—it meant some calves would be born after the planned release from the pasture into the wild—but perhaps it would turn out to be for the best.

Later births could slow the animals down even more, making them less likely to roam too far.

Regardless, we started the final preparations to release the animals on schedule in late July. In consideration of Wes Olson's target concept—guiding the bison to explore nearby areas first, then slowly exposing them to more distant areas—I asked everyone to put in extra time to build a release corridor. My vision was a kilometre-long path bounded by berms of dead wood, with a trail of bison manure to lead the animals through the middle. If it worked, this setup of obstacle and scent would guide the

animals to the closest meadow upstream from the pasture, where they could explore within striking distance of home and then return for hay and salt.

If any of my colleagues doubted my vision of how the release would unfold, they remained silent during the building of the later-infamous release corridor. Everyone pitched in to cut and haul dead wood and wheelbarrow over hundreds of loads of dung from Manure Mountain. A few of us even carried a generator and an electric jack-hammer to chip and smooth a rock step that we thought might be too awkward for bison.

We met our goal and were ready for the end of July. Only half of that year's calves had been born, but everything else had lined up according to plan. We'd release thirty-one animals instead of thirty-six, at a time when the grass was lush and growing at its peak. The yet-to-be-born calves would keep the bison closer to the home pasture. Or so we thought.

Everything was ready, except for one thing: blessing the animals for the upcoming transition to the wild. The public had wrongly assumed the reintroduction was over when we'd flown the animals into the backcountry a year and a half before. In truth, we were only getting started. Indigenous Nations understood the seriousness of what was to come and wanted to acknowledge it. The day before we released the animals from the pasture where we had been ranching them, a helicopter with Elders from the Treaty 7 and Métis Nations arrived.

I was intrigued but stayed out of the way as the special day unfolded. It was different from the Elk Island ceremony: more intimate and low-key. There were no loudspeakers, no teepees, headdresses, or speeches. Different groups of Elders went off to different areas outside the holding pasture and uttered prayers and sang songs while the bison enjoyed their last meal of hay behind the fence. I watched, absorbed but also a little ashamed that, given all we had invested, my own culture did nothing to acknowledge this big moment.

The best we could do was fly in my boss, Bill, and his boss the next morning, after the Elders had left. While the animals slept in a lower corner of the pasture, the three of us snuck across an upper portion and cut the fence. We symbolically took turns with the sharp pliers, snipping through the wire and pulling the mesh back, creating a five-metre-wide hole into the release corridor.

"We're out of here," Bill said, after they had waited two hours and none of the animals so much as stirred. He held a fist to his ear with his thumb and pinky poking out. "Call me tomorrow," he said as the helicopter started up. "Let me know how it goes."

I nodded and smiled. Little did either of us know we were about to get buffaloed.

10

BE ADAPTABLE

THE BISON FOUND the hole we had cut in the fence in the middle of the night. Where and how the animals moved in the first hours of their release demonstrated that our immediate concerns had been unfounded. The bison were not unprepared for being in the mountains; it was we who had assumed—wrongly—that they were not ready.

I had a fitful sleep after cutting the fence and was up early, padding out to the porch in the pre-dawn light. Flicking on the receiver used to track the animals, I checked for signals from the radio collars. No response. A wave of panic washed over me. If the bison were anywhere in the Panther Valley, beeps should have resonated loud and clear.

Adam, stirring awake on the cabin couch, heard me curse as I fiddled with the dials.

"What's going on?" The sleepy look on his face cleared when I told him. By the time the rest of the team was up, Adam had pulled on his clothes and run off.

"I'll look for hints," he called over his shoulder as he headed to the release corridor. "I'll be back in half an hour."

The others gulped their coffee and, after a few bites of breakfast, packed their lunches. No one knew what to expect when Adam reported back, but we were certain it would be a long day. When we heard him clomp up the deck stairs, we all became silent.

"They found the hole, all right," he gasped, still catching his breath, "and they followed the corridor. But they didn't go into the meadow at the end like we intended. They veered north instead of west. I doubt they even saw all that open country available to them." He shook his head and took a deep breath. "They pushed through that dark thicket of trees instead." He grabbed his wrist with his other hand. "Broke branches as thick as my arm! They just kept going when they got to that little canyon." His hand pantomimed their slips and slides as they descended one side and climbed up the other.

I knew exactly where Adam was referring to. I had pointed out that canyon to Wes when I was trying to establish how bison might use the landscape. "Not a chance!" Wes had replied. The bison, however, wasted no time defying those expectations.

"I didn't go any farther," Adam said. "I could see from the tracks there's no hope of catching up." He raised a hand and floated it upward. "They're climbing. Climbing and climbing. Prints gouged up the slope. It's like they're possessed!"

Everyone looked at me expectantly when Adam was done. I stood speechless.

How could we be so naive? I wanted to ask. How could I be so arrogant and assume that what happened in Jasper would not happen here?

I felt like throwing up my hands and screaming but instead recognized this as a leadership moment. Each face that looked back at me was committed to the project in part due to the pep talks I often delivered. I reminded myself of what we were trying to accomplish—to return a species that belonged and deserved to be here. I had to channel the art of the possible and not the fear of failure. After buoying myself up with my own inner monologue, I proceeded in the way leaders often do in moments of great indecision: I stalled.

It was 7:30 AM, time for the morning radio call to the main dispatch office in Banff. As we all gathered, I clicked on my walkie-talkie and gave an update on the situation. A firefighter, listening from town, jumped in at the end of the exchange.

"We have a helicopter sitting here on standby. Sounds like you could use it," she offered. "It's high fire hazard, but we won't do any smoke flights until it heats up in a few hours. The machine is already paid for. You can borrow it. Just have it back here by ten AM."

Twenty minutes later, the helicopter arrived outside our door. Saundi and Pete White, who had helped us with many of the preparations, loaded the dart gun and other capture gear into the rear hatch and climbed aboard; I sat in the front. Adam and Dillon stayed behind to spread hay, mend the hole in the fence, and open all the gates in case some animals returned to the pasture.

As we took off, I directed the pilot to fly high while I toggled through the radio collar frequencies again. Faint blips sounded as we climbed above the surrounding ridges. I exhaled in relief.

"They're still in the park," I announced over the helicopter intercom. "Up this narrow valley." I pointed and the pilot swung north.

The radio collar signals plinked loudly as we neared the top of the valley. The trees gave way to a triad of alpine basins, with blue lakes that twinkled like jewels in each one. We saw no animals in the open meadows of ground-hugging plants and shrubs, but the signals were strong; the needle of my radio receiver bounced hard. Bewildered, I glanced at the pilot. He shrugged back.

And that was when I saw them, through the window behind him.

"Look up," I blurted. "Way up!"

The pilot spun the helicopter around so everyone could see. The bison came into view like apparitions along a high ridgetop, a line of buffalo-robed mountaineers closing in on a summit. The movement of something red caught my eye. I was gobsmacked. A calf, born in the pasture just two days before, was running across the rocky heights, chasing its mother and the other calves with the surefootedness of a child in a schoolyard.

"What the—?" Pete mumbled. Saundi looked stunned.

The pilot stuttered. "They're like . . ." He struggled to find words. "They're like oversized mountain goats. Only brown instead of white!"

I asked for us to land on a nearby ridge so we could watch from a distance and figure out what was going on.

When we stepped down onto the ground a few minutes later, we were still mesmerized by the scene of the marching animals. Smoke from a distant forest fire gave

a pink cast to the panorama, making our view seem like a dream. Peering through my binoculars, I counted thirty-one animals. Another wave of relief swept over me. Only two bulls were missing; all the females and their calves and yearlings were present, uncharacteristically bunched together. They looked and behaved like a herd, free of the usual jockeying and pecking-order behaviour. Without fences and tight quarters pressuring them, the animals moved more as a single entity, as if being wild allowed them to be more themselves.

"They're chewing their cud!" Saundi exclaimed, squinting through the spotting scope she had set up. About half the group lay on a gravel terrace while the others milled around them like a tight-knit family, weaving between broken cliffs. Above them, serrated rock slabs jutted into the sky like a dragon's back. "Some of them are dozing!" Saundi said as she scanned the animals. "They're sleeping on top of the world."

"It's as if they've spent their entire lives in high-altitude places," commented the pilot, stepping out of the cockpit.

"Good thing we put in all that work to build the release corridor," Pete joked after we'd watched the animals for a few more minutes. He had logged a lot of days on the chainsaw.

"Yeah," Saundi added, "good thing we pounded in all those extra fence posts and included a hill in the pasture." She rolled her eyes. "So they could adjust to the steep ground!"

"Sorry," I gestured. It was me who had insisted they do all that extra work, based on the bison behaviour I'd anticipated. Clearly I did not know enough. We were all learning in real time now.

"What *are* they doing up there?" Pete finally asked.

"Maybe escaping the heat," I suggested. "Or the bugs."

"Or having a good look around," said Saundi. "Orienting to a new area."

Pete, who took a turn behind the scope, offered the most likely explanation.

"Maybe food?" he suggested. "They're eating something. Green sprigs growing between the rocks."

I gently shouldered him aside and adjusted the focus ring. Indeed, they were feeding on something growing between the frost- and wind-shattered rocks. Was it possible, I wondered—had they instinctively known to go up for the most palatable food? Other big mountain herbivores, like elk and bighorn sheep, know to follow the melting snows for the best sprigs of new plant growth by climbing up. But how had bison from the flatlands of Elk Island known? What was speaking to them that we couldn't see or comprehend?

Shivers ran down my spine as the pilot pointed at his watch. Knowing I couldn't answer such questions yet, I was nonetheless excited by what they suggested: that the bison were motivated in ways we had yet to understand.

11

DO NOT OVERREACT

ONE OF THE TWO missing bulls showed up at the park boundary the next day; the other one appeared a few days later. They were on separate journeys, but each incited a chase as they left the park.

Bull Number 5, the Pensive One, who often stood looking out of the pasture, didn't waste time pursuing his dream of freedom. The drift fence across the Red Deer Valley, at the eastern park boundary, was in bison-deflection mode, but it made no difference. He simply jumped it after pacing back and forth a few times. Day two of the release and he had already called our bluff.

I managed to get in front of him on horseback a few times and tried to herd him back, but he was determined—driven by instinct. Was it the search for receptive females that fuelled him? Was he escaping competing bulls? Was he drawn to return to Elk Island? One thing I knew for sure: he was not motivated by food. He plowed his way through the Ya Ha Tinda area without even pausing for a mouthful of grass.

The situation became serious when the collar locations, which were downloaded to the web every twenty-four hours, showed him beelining out of the mountains. He had roamed onto leased Alberta public land where ranchers graze their cattle, about thirty kilometres east of the park. I was reminded of what was at the root of the ranchers' concerns, and consequently, what prevented the Alberta government from treating plains bison as wildlife, like other animals such as deer and elk. They worried that bison, being wild bovids, could carry diseases that could infect their livestock, which are also bovids. We minimized that risk by testing and quarantining the bison repeatedly, committing to act quickly if any disease were detected and to deal with any bison that left the park. With Number 5 on a fast trajectory to enter the town of Sundre the next day, it was time to honour that commitment.

From my Banff station, I called a local helicopter company and arranged for a machine and two of our best shooters to look for him. The speed at which he was moving, and the fact that a lightning storm sweeping through the area was starting many forest fires, meant we didn't have the resources to recapture him. I sat by the VHF radio in the meeting room that doubled as a command centre for incidents like this—a classic soulless government office, with particle-board furniture and fluorescent lights. The setting was so different from what occupied the dreams of Number 5, dreams he pursued in earnest right up until he was found and put down.

"Bison Gunned Down!" screamed the front-page headline of a Calgary newspaper the next morning. "Protected?" asked another. Hundreds of posts on social media were just as unkind.

We had barely regained our bearings when, a day later, the signals of bull Number 19 also registered outside of the park. By then, my bosses were done with negative media.

"Capture him alive," came the orders from above, "no matter the cost."

Number 19 had also left the park via the Red Deer Valley, but instead of jumping the drift fence, he skirted it, climbing high and around an end to pick his way through some steep rocks. He travelled north from there, eventually getting to the Clearwater Valley, which he followed east out of the mountains. That was where we caught up to him, near a gravel forestry road, already in a grazing allotment, not bothering the nearby domestic cows.

No one would have known he was there if not for the radio collar. The bushy, swampy countryside is mostly forested and has poor sightlines. We needed several passes over in the helicopter before we finally found him, at the edge of a series of meadows.

We landed in a nearby opening and set off on foot. The plan was that we would dart him with a tranquilizer gun, but he eluded us. We quickly transitioned to Plan B: put him on the ground with a net gun. This required specialized equipment and the skills of the capture crew I had hired—a sixty-year-old ace pilot and his unassuming, rough-and-tumble son. For the next fifteen minutes, we watched as the father flew the helicopter and his son stood on an outside skid, with the doors removed and a rope tied around his waist. He carried what looked like a cannon tucked under his arm. A minute later, he raised it to his shoulder, and the net gun fired with a crack and puff of smoke as the rotors dipped into a hidden meadow behind the treetops.

"Come get 'im," drawled the dad over the radio. "Buffalo down."

We ran to where Number 19 lay tangled in the net and jabbed him with a syringe of tranquilizer. Once he was asleep, we slipped hobbles over his hooves and used ropes and pulleys to winch him into a bag of heavy canvas and webbing. The helicopter flew overhead, this time with the son, also a gifted pilot, at the controls. We hooked the bull to the longline that hung from the machine's belly.

I had arranged for a truck and trailer to be positioned at the nearby gravel road. With some fancy flying, the pilot swung the sleeping bison inside the door of the covered horse trailer and gently placed him on the floor. It saved us from winching and dragging him in. We easily slipped off the carry bag, hobbles, and blindfold, and injected another drug to reverse the tranquilizer.

It was too risky to return him to the Banff backcountry, where he might have led others from the park. Instead, we drove Number 19 to Waterton Lakes National Park in southern Alberta and added him to the small captive herd that lives there. The fenced enclosure was a little larger than the pasture he had called home for the past year and a half in the Panther Valley. After he'd experienced only a few days of freedom, I felt bad for delivering him there.

It was a small consolation a year later when genetic testing we completed for other purposes confirmed Number 19 as the dominant bull while the herd was in the Banff backcountry. He had sired most of the calves born that second spring, and although he would not run wild himself, his genes were well on their way.

———

THE CAPTURE OPERATION for Number 19 cost more than twenty thousand dollars, not including employee and contractor time. I cringed to think what it would cost, and the effort it would take, if the whole herd followed suit. I was still haunted by the Jasper failure.

After conferring with Bill Hunt at our next meeting, I decided to reduce the risk of a mass exodus by extending the fence that one of the bulls had skirted. We knew from our fence research that we couldn't increase the height—that would compromise permeability for other wildlife—but we could lengthen it up the rocky slopes to the base of the cliffs. This would block the terrain the experts had previously assured us bison would never use, but those assumptions had already been proven wrong.

We had thought we were done with fencing, but now we donned gloves and hard hats again. Everyone knew the tasks to be completed, making it easier to direct a gang of summer students we recruited to help. We put in long days, working into the evenings, and were finished in a week.

We wanted to get the job done quickly because—as far as we knew—the other bison could follow one of those bulls and leave the park at any moment. Fortunately, that didn't happen. Ultimately, it was seven months before the other bison visited that fence, and they certainly never went upslope far enough to justify our extending it. In fact, we removed the extensions five years later. That instalment turned out to be a ninety-thousand-dollar overreaction.

In hindsight, all our responses to those initial wayward bulls were overreactions. A bison or two leaving the park in late summer became an annual occurrence, something to be expected. A year after Numbers 5 and 19 left the park, for example, Number 2, another bull, wandered through acreages around the town of Sundre for days. We finally darted him on the lawn of a ranchette. Despite being out for days, he had caused no problems. Similarly, there was the young male that poked his head into horse camps outside the park in the Clearwater Valley, who I eventually found and struggled to shoot. Although backcountry campers were surprised to see him, they were also saddened at the outcome. The newspapers did not report either incident.

Bison on the landscape were becoming normalized.

12

HONOUR HISTORY

U NLIKE THE TWO wandering bulls, the rest of the
bison stayed within two square kilometres of the
ridges where we found them after the release. They
remained in a cluster of basins surrounded by rock walls
for the next month and a half. Then, in late September,
with the autumn frosts curing the alpine tundra and turning the willow bushes yellow, the animals decamped. They
returned to the pasture, sniffed around its open gates for a
few hours, then continued east. They ambled for thirteen
kilometres, through a gap of mature forest downstream of
the pasture to where the surrounding ridges leaned back
and gave way to grassy hills, plateaus, and knolls.

This was the area of the park that Wes had liked. Shel-
tered from winter storms, which stall on the ridges to the
east and west, it occupies a dual rain shadow, an anomaly
of arid grasslands where warm chinook winds touch down
and melt the bit of snow that falls in skirts around the grey
peaks. Littered with the old antlers of past overwintering
elk, this was where the bison stopped and explored a giant

doughnut of meadows for the next few months. As when I rode through the area with Wes two years earlier, the trails held the recent markings of wolves. There were even more fresh tracks and scat now, a sure sign the ancient dance of predator and prey had begun again.

By the end of February, we had enough uploaded information from the radio collars tracking the bison to see a pattern: The bison explored in surges—like a flock of migrating birds who pause at known stopovers with good food, to tank up before pushing farther into places that might have marginal resources. Academics who study such animal learning and migration call these areas "encampments"—pads of certainty from which animals launch riskier explorations into the unknown. The

Banff bison "encamped" in that lower Panther area for six months, discovering every nook and cranny.

The animals surged again early the next spring, slipping north over a low treed pass and entering what became their favourite valley—the Red Deer—for the first time. They encamped again at some meadows formed by a tributary that often floods and keeps the trees at bay, making room for grass. Limited by the gravelly soil, the grass that grows there is spindly, not lush enough to hold bison for any length of time, and after a few weeks, the animals explored on. Watching their locations on our screens, we were reminded again of the question that faced us on that ridge the morning after the release: What drives them to explore?

I knew one thing: It was not the recent past informing them. Being a reintroduced population, the animals had no prior knowledge of the area. The fact that this didn't lead to mistakes provided a clue. Despite a plethora of cliffs, canyons, and dead-end valleys in the area, the animals did not get entangled or turned around once. How did they know to avoid such places?

The mystery persisted as the bison climbed high again that second summer. They went up a side valley, like they had the previous year, but this year, it was even better habitat for them—more bowls and slopes covered with young and palatable plants. They followed sheep and elk trails linking it all: a high-elevation network that connected emerald basins to greening plateaus, bounded by towering rock walls and soaring peaks. The bison spent two months encamped there, their big, brown bodies strangely oversized amid the delicate alpine plants. On a few hot days, we found them lying on lingering snow patches, escaping the worst of the heat. Bewildered, I asked myself if it could be random. Were they haphazardly bouncing around like one of those robot vacuum cleaners bumping into furniture, learning the layout of a house? Was that how they discovered such refuges from the heat?

Further clues were revealed that fall, when the bison came down from the high country and discovered what became their favourite meadow—a swath of rich grass and deep soil that had slumped off an unglaciated mountain thousands of years earlier into the Red Deer Valley. Having escaped the scraping and scouring ice that had shaped the area, the unique meadow hosted a diversity of grasses and forbs; it was like a mini Ya Ha Tinda grassland.

How the bison arrived was instructive. I happened to be staying in a nearby cabin. The meadow lies to the side of the main valley and is guarded by steep mountains, thick forests, and a river canyon on all sides but one. The bison beelined in via the only available route, as if they knew it was there all along.

That meadow became the animals' new hub of activity—the place where they spent their second winter, where they calved the next spring, and where they launched from to go high again the following summer. A network of bison trails began to radiate out from it like spokes on a wheel, linking to all other places they favoured.

Our team explored these trails on our multiday field trips, poking into areas where the collar downloads indicated the animals spent time. We did this after working hours: inquisitive evening ambles after days of checking drift fences or collecting samples of plants the bison ate, or their dung. We did it to satisfy our curiosity about the animals and to get valuable insight into their world.

These walks through the bison landscape deepened our sense of the place, revealing hidden seeps, tiny meadows, and shortcuts that, despite having ridden past them on nearby horse trails countless times, we never knew existed. Something else began happening: As the bison used the trails more, their hooves churned deeper into the soil, revealing another story.

It was Pete who noticed the beginnings of what became a trend. He came back from a shift with a bag of ancient bison bones he'd found at various places along one of the new bison trails.

"I've never seen anything like it," he said, handing over the bag of bones to me at the office. The next week, one of the Parks archaeologists stumbled on more old bison bones in the holding pasture where the reintroduced animals had rolled and pawed, exposing the ground. Wondering if something was up, I visited some of the freshly created trails in the Red Deer Valley that Pete had described. No sooner did I get there than an old bison skull presented itself, poking out of the ground.

"What is going on?" I wondered aloud as I tugged at the old horn. With the sun sliding off the surrounding peaks, a piece of the puzzle dropped into place. The bison we had reintroduced, I realized, were using signs left by their ancestors to find their way across the land now.

I would not have been as open to such a possibility had Leanne and I not lived among caribou and experienced the wonders of herd consciousness. Were these bison tapping into something similar to talk to their long-dead ancestors? Were they bending time?

This was big, ideological stuff to grapple with. As with the caribou, I was challenged to meld this experience with my educational background in science. It was more of an Indigenous way of thinking—imagining everything being alive and connected across time.

I wandered back to the cabin, opened the door, and paused on the porch to bid the day good night. An Indigenous greeting often shared before and after our meetings, which acknowledged that everything is alive and connected, seemed appropriate for how I felt.

"All my relations," I uttered as I shut the door and settled in for the night.

13

BE WHERE YOU BELONG

SHARED MY NEW realization about the bones with discretion, thinking some people might label me a quack. One person I did confide in was Kent Prior, one of my bosses at Parks Canada's national office. Kent oversaw the national Conservation and Restoration Program, which funded the project, and when I invited him for a four-day backpack trip through the bison zone to see how we had spent the money, he accepted. Despite driving a desk at work, he maintained an active lifestyle, and in the field he had no problems keeping up. After a few days, it became obvious to me that we were cut from the same cloth. On our last day I decided to take a chance and unfurled the story about the bones.

He listened as he stood in a freshly excavated wallow—a depression of bare soil that bison dig and roll in to rid themselves of loose fur and bothersome bugs. It was our final day, and I knew this was the last place we would see major bison activity before we exited the park. Kent seemed distracted, glancing down frequently as I finished describing what we'd found.

"Like this?" he asked, reaching to pick up a very worn-looking bison knuckle bone. A huge smile spread across his grizzled face.

When I returned to Banff, I boxed up that old bone with the others Pete and I and the archaeologist had collected and sent them off to be analyzed by an archaeological lab in Eastern Canada. I checked off two analyses on the order sheet to be completed: radiocarbon dating and isotopic testing of the collagen.

The carbon dating results came back within the range of what we already knew about bison bones in the park; they were between 270 and 2,100 years old. However, the collagen revealed something new.

There had been a long-standing question about the bison in the area before they went locally extinct: Were they year-round residents or seasonal migrants? If the answer was "residents," it would confirm what we were trying to restore: a plains bison population that lived in the mountains throughout the year. But if the answer was "migrants," it suggested we were trying to restore a state that had never existed here, likely because the mountains didn't provide all that bison needed to survive. It was an argument the cattle lobby liked to use to discount our efforts.

The ancient bones unearthed by the reintroduced bison validated what we were doing. Using differing atomic weights, technicians were able to discern the C_3 to C_4

isotopic ratio in the collagen of each bone, and thereby extrapolate where the ancient bison had primarily spent their time. High C_3 content indicates a diet rich in cool grasses, which grow in the foothills and mountains and leave a strong signature in the bone collagen due to their unique pathway of photosynthesis. High C_4 content points to a diet of warm grasses, from which we would infer that the ancient bison travelled or "migrated" to areas where such grasses primarily grow, namely on the prairies.

All the ancient samples unearthed by the reintroduced bison showed a high C_3 to C_4 isotopic ratio, strongly supporting the theory that ancient bison had lived year-round in the mountains as residents. The bison themselves had confirmed that the project was not misguided. Together, we and the bison were restoring a corner of the world to what it was before humans nearly wiped the buffalo out.

14

KNOW THY ENEMY

I WAS ENCOURAGED TO KNOW that bison had lived year-round in the mountains around Banff, and that we were replicating a situation that naturally existed. This was even more special given that the Banff plains bison herd would be only the second to roam with both its native predators—grizzly bears and wolves—within its historic range. There was only one other population for which this was true: the herd in Yellowstone National Park.

Predators are big shapers of evolution. Aldo Leopold, the well-known ecologist and father of modern conservation, famously asked: "Where would the deer be without the wolf to have whittled its leg?" In other words, without predators, slow and sick prey pass on their weak genes, reducing the species' fitness over time. The success of past predators taking down bison leads to a positive outcome for faster, stronger bison survivors today.

Despite this, a retired university professor who was an expert on wolf and bison had warned that our plan to reintroduce bison to Banff National Park was destined for failure. For decades, he had studied how wood bison—a close cousin to plains bison—had interacted with wolves in the hinterlands of northern Alberta's Wood Buffalo National Park. He predicted the Banff bison herd would also be chased by wolves and run far outside the park.

"Good luck keeping them in Banff," he said when Parks Canada first floated the idea of the reintroduction in 2012. "That park is tiny compared to Wood Buffalo—a fortieth the size. I've seen wolves push bison over one hundred kilometres!"

I was heartened that it had taken wolves in the Yukon twenty-five years to learn how to take down wood bison after they were reintroduced there in the 1980s. Maybe time was on our side.

As Wes had said during our backcountry horse trip through Banff Park in the fall of 2015, bison are more likely to stand their ground than elk or other large prey. If a predator happens to get its fangs or claws into a bison's thick hide, the battle is not over. Nearby herd mates often help. Their powerful short legs and low centre of gravity make it difficult to put a bison on the ground. Even then, it will continue to fight, and predators risk a lethal blow from lashing hooves and swinging horns until a bison's remarkable stamina runs out.

The animals in Yellowstone provide an example of bison's incredible resilience to predation. Approximately two thousand bison lived in the park when wolves were reintroduced in the early 1990s. Surprisingly, the number of bison increased afterward, tripling to more than six thousand by 2023. By comparison, elk fared poorly following the wolf reintroduction, declining from ten thousand to thirty-five hundred animals in a decade. The few bison that wolves kill tend to die in late winter, when they succumb to being run through a deep snowpack while the lighter, splay-footed wolves float on top of a supportive crust.

All this to say: We were unable to predict how these bison from Elk Island, which had never seen a wolf or grizzly bear, would interact with their predators in Banff.

If money and effort were unlimited and animal welfare were assured, we could radio-collar every bison and wolf to tell us exactly how the two species were responding to each other. This was not our reality. We already had our hands full trying to keep enough bison radio-collared to monitor them; adding wolves made things even trickier. No sooner would we deploy a collar on a wolf than it would end up in a trapper's snare just outside the park. One trapper operating just beyond the boundary was particularly effective, single-handedly killing more than half of the seventy-two radio-collared wolves that have died in the area over the past three decades.

Prior to bringing the bison to Banff, I had taken another chance and hired the father-and-son heli-net-gunning team, and they caught two wolves for us, attached radio collars, and released them in the same area where the bison were destined to go. We gathered a lot of information about those wolves before and just after we released the bison. But after only six months, the wolves left the park and were lured into that infamous trapline, where they met their ends.

The locations of those two wolves prior to their untimely deaths showed that they were frequently investigating the bison: about once a week after the bison were released. Although their collars registered in the same meadows as the bison—often within a few hundred metres—the bison collars didn't budge. It was always the wolves that moved, often far away, to check out other sections of their vast territories before returning about a week later, when the whole scenario would play out again.

Our team periodically saw wolves or bison on their own, but we never saw them together. A network of electronic eyes, deployed at mineral licks, alpine passes, rub trees, and other spots wildlife frequented caught them mingling periodically. These remote cameras were deployed by Parks Canada's ecological monitoring crew about every ten square kilometres throughout the park. Originally developed by third-party manufacturers for hunting, the cameras are an efficient, non-invasive, low-cost method for monitoring wildlife across vast areas. Trends are extracted from the collected imagery and used to tweak park management, but the cameras also capture moments that would otherwise be missed: scenarios like two wolves following a group of bison only to run past the same camera in the opposite direction a few minutes later, chased by two agitated bulls. Another camera a few kilometres farther upstream captured images of dozens of bison calmly grazing as a wolf circulated in their midst for hours.

Other images showed the Banff bison fitting in among their fellow wild animals: a band of bighorn sheep shared a mineral lick with the herd; a family of lynx crossed the

Panther River within moments of the bison crossing; and countless cougars, coyotes, deer, elk, and bears used the same trails as the reintroduced bison. One camera captured a heart-wrenching moment when a broken-legged bison calf hobbled behind its mother as a drooling wolf watched in the background. We never saw that calf again. Although sad, this was rare and represented the only bison injury or death we witnessed in seven years.

Given the pressures to keep the reintroduced bison in the park, I was relieved that the chase sequences the professor had forewarned us about did not occur. For whatever reason, the resident wolves, bison, and grizzly bears were taking their time getting reacquainted. Things were bound to change once they did.

15

MAKE YOUR IDEA THEIR IDEA

WHETHER IT WAS their ancestors, wolves, or something else that set them off is unknown, but partway through their first full summer in the wild, the Banff bison herd left the Red Deer Valley, went up a tributary valley, headed over a pass, and marched out of the park.

"Get your backpack, a couple of BB pistols, and the flags," I barked at Adam as he came into the office that morning. "Meet me outside in fifteen minutes. The helicopter is on its way."

We spotted the animals in a wet meadow six hundred metres beyond the park boundary. Half of them were bedded; the others milled and fed in the soggy, football-field-sized patch of moss and grass. Pools of standing water reflected shards of blue-and-white sky up at us as we flew by. I counted the brown lumps as we passed high overhead. There were fifty—most of the animals we had released a year ago, plus the calves born in the wild that spring. A few missing bulls were off alone, someplace else.

I asked the pilot to circle as I cobbled together a plan. I had been there, on the ground, many years before, when I was a warden patrolling for poachers who might shoot a bighorn ram on the wrong side of the park boundary. From the helicopter, I scanned the scree slopes and rocky peaks above the green pass. The meadow sat like an open hand balanced on top of the slopes, outstretched fingers disappearing into the pine and spruce forests that blanketed much of what was within view. There were no rams in sight, and, thankfully, no people to see what happened next.

Herding, which is defined as moving animals in the direction of choice, is commonly used by farmers and ranchers to usher livestock between feeding and handling areas. We needed a similar tactic, because the perimeter of our system of pastures—the bison zone inside the park—was porous. Large stretches of the park boundary were impassable due to cliffs and other rugged terrain, but viable exits remained. We had blocked many, but there were quite a few mountain passes, like this one, by which bison could leave the park. If they chose to do so, our only option was to herd them back.

Not many people herd bison, and particularly not in places where driving a vehicle is impossible. The only exceptions I'd discovered in researching the subject were on the vast bison ranches owned by Ted Turner, the American founder of the CNN broadcasting empire. Until recently, he was the biggest landowner in the United

States and had converted most of his land into huge bison ranches. Many of the internal fences had been removed, and the bison were herded using a technique called low-stress stockmanship, often done on horseback.

To learn more, in 2016, before our reintroduction began, our small project team had travelled to one of Turner's ranches in the Snowcrest Range of Montana to ride and learn from their cowboys. The technique was very different from the usual ram-and-slam-them approach of most livestock operators. Developed by American Bud Williams in the 1980s as he pushed tens of thousands of cattle through a Canadian feedlot over seven years, low-stress stockmanship was based on the animals' natural, not forced, responses to pressure. People took notice when the animals he handled came through the facility having lost significantly less weight than usual. Profits soared. His techniques have been used throughout the world to move wild burros, feral horses, reindeer, grizzly bears, and now, bison.

The root of the low-stress approach is the idea that every animal is surrounded by two bubbles: an outer pressure zone, defined by the distance at which an animal acknowledges a person's presence; and an inner flight zone, the distance at which that person's presence is no longer tolerated and the animal flees. The area between the two distances is where the herder operates, exerting pressure and releasing it, rewarding the subject animal.

Now, I tried to tap into all our herding training and research and adapt it to an airborne approach as our helicopter hovered over our boundary-crossing herd.

"If the bison start running, we're doing it wrong," I said as the pilot dropped our elevation and we approached the animals. I raised my hand to indicate he should go slowly, citing two of the low-pressure technique's basic principles: 1) Keep the animals in a normal state of mind, and 2) Make our idea their idea.

Understanding where a person should be in relation to the animals' two pressure zones comes from watching the animals themselves.

"That's far enough," I said as some of the sleeping animals lifted their heads. Others stood and relieved themselves, yawning and stretching like oversized dogs. "Hover sideways, please," I instructed the pilot.

The pilot obliged, working the foot pedals like an expert piano player, following the cues of the animals. We drifted back and forth, gently nudging the bison toward the park.

"A little lower," I suggested. Then, as his hand pushed down on the collective lever, I countered by raising my own hand in the air. "Up ever so slightly." Ripples streamed across a nearby pond and leaves blew off bushes from the wind of the rotors. "Easy," I said as the animals closest to us squinted in our direction. Slowly, some of them took a step. Then another. A few seconds later, the bison trickled off the pass. They walked two to three abreast until they hit the horse trail to the Red Deer Valley, at which point they stretched out—one behind the other—travelling in single file. I watched in disbelief. Similar attempts to turn the wayward bulls around the previous summer had failed, but this time it was working.

"Good movement attracts good movement!" I shouted into the microphone, quoting our stockmanship teacher.

"They *want* to be in a herd," Adam quoted back euphorically.

We continued behind the herd, politely hovering, pushing the animals for another kilometre before they disappeared into the trees below the pass.

"Put us down there," I said, pointing at the last opening where the pilot could land as the forest closed in. "We'll continue on foot from that spot."

Adam climbed out after we landed, smiling, visibly eager to push the animals deeper into the park. The pilot shut down and settled in to wait as I opened the rear hatch and grabbed the BB pistols and flags.

"Here," I said, shoving one of each into Adam's hands, and we both headed off.

We followed the animals' fresh tracks another kilometre to where they left the trail and veered up a grassy avalanche path.

"They're up here," I gasped, fiddling with the portable receiver for a hint from the distant radio collars. I was struggling to keep up with Adam's mountaineering pace as we climbed, and happy for an excuse to slow down.

"Hang on," I gasped again, spinning the antennas with a raised hand. Beeps sounded and I took another step upward. "They're close."

We found them on a bench where the avalanches had piled a berm of broken timber on the edge of a lush, green meadow. Clambering over the fallen trunks, we could see the animals eating the rich grass. We tried all the stockmanship tricks to get them moving again—walking in and out at a forty-five-degree angle, approaching and retreating from straight on, and rocking back and forth on the spot—but the animals wouldn't stir.

None of us were giving an inch in our little standoff. As I waited, I recognized many members of the herd by their green ear tags. Unsurprisingly, Number 6 was out front, followed by Numbers 10 and 11. The other dominant cows were close behind. Numbers 7 and 17 were at the back as usual. Bulls Number 2 and 4 stood in the middle of the herd, looking as relaxed as ever, rubbing shoulders with each other as they chowed down mouthfuls of grass. But they had all changed visibly. Twelve months in the wild had taken its toll. Their faces were marked with bumps and scrapes, horns were broken and bumped off, and their once plump, hay-fed bodies were skinny and hard. Their ribs and hips showed beneath dull coats matted with small branches and mud.

It was Adam who caught their attention. Reaching into his backpack, he unfurled the square polypropylene flag, fashioned from a tattered tarp and a broken axe handle. Half of the animals turned to face us. Seeing how well this worked, I decided to increase our presence, and hence the pressure, even further. I pulled out my BB pistol and fired it into the air. It was too much. The rest of the animals' heads lifted, and the whole herd bolted away.

We scrambled into an old burn scar on the other side of the avalanche path but were unable to keep up. Out of breath, I signalled for Adam to stop. The snap of breaking branches faded in the distance ahead of us.

"Let's head back," I suggested. A carpet of purple fire-weed closed in behind the stampede of panicked animals.

The herd had gone five kilometres deeper into the park by the time we saw them from the helicopter again a half-hour later. They were filing through a large area of shrubs as we flew overhead. We gently encouraged them for a few minutes, but soon I saw no need to pressure them further. They were strung out in single file again, heads to tails. Something about how they moved—legs swinging and high-stepping, pointed on a down-valley trajectory— told me we were done. Their mindset had switched from exploring the unknown to seeking familiar territory. I asked the pilot to take us back to Banff.

It was a good decision. After returning the BB pistols and flags to their respective cupboards at the office, I turned on my computer and called up the GPS-collar platform. A few collar locations downloaded, and sure enough, the bison were back at one of their favourite meadows in the Red Deer Valley, fifteen kilometres inside the park.

The animals never tried to leave the park via that pass again, and we used the technique to resolve similar threats half a dozen times in the following years.

But on the day of that first successful helicopter roundup, I glanced at the collar downloads one last time before putting my computer to sleep. I couldn't help smiling as I trundled down the hall and emerged into the office parking lot.

It had been a great day. We had made our idea their idea.

16

TRACK YOUR PROGRESS

A YEAR AFTER THE bison's release in 2018, the radio collars we'd initially put on the animals began to fall off. Fewer marked animals meant fewer ways to track the project's progress and compromised our ability to respond to animals leaving the park. How would we know when herding was necessary, or which fences needed to be deployed?

In yet another instance of unintended consequences, this problem of collars falling off was the result of an abundance of caution when the animals had been in the holding pasture. A best practice when radio-collaring large animals was to include a piece of canvas fire hose, as a weaker link, in the collar's bolted closure. This acted as a fail-safe backup to the factory-made blow-off mechanism—which was meant to cause the collars to fall off when the battery died, but often failed. The risk of the hose inserts, though, was that they rotted, rubbed, or wore through prematurely.

Years of trial and error with other animals suggested optimal numbers of canvas layers to insert: A single layer worked for herbivores up to the size of an elk, and two layers worked well for moose and bears. We guessed three would be ideal for bison, but we were wrong. The animals' extensive rolling and lying in dirt, tussling, and rubbing on trees, rocks, and each other was tougher on the canvas than we expected.

Radio-collaring bison in a backcountry pasture was difficult enough; the thought of doing it in the wild overwhelmed us. In the pasture we had advantages—a fence to enclose the animals, a snowmobile or quad to travel after them, hay to bait them with, and a nearby wood stove to thaw out the dart gun and keep the sedative drugs warm—but it still took our team and a wildlife veterinarian two weeks to complete the job. How would we accomplish this in the wild? I called the father-and-son helicopter capture team once again.

"We're really busy with caribou work," they said. I booked an open slot they had later that winter.

The Wild West duo touched down at the Banff helipad early one January morning. We reviewed the map together, and I shared radio frequencies for the remaining four collared bison so they could easily locate the herd.

"Don't forget to remove the magnets that temporarily disable these," said Saundi as she handed over two new collars for this initial attempt. Father and son fired up the helicopter.

"Try to get some blood for disease screening," I added, shouting to the son as he pulled the front door closed and handing him a bag of tubes and syringes through the window. He nodded and tossed them under his seat, along with the wad of park aircraft landing permits I shoved in. He nodded again, and they took off.

They were back two hours later, all smiles.

"Got two collars on," the father reported after they touched down. He kept the helicopter running as his son returned the bag of tubes and syringes—they were empty, but his coveralls were stained with fresh patches of blood.

"We tried," he said, throwing up his hands apologetically as a tuft of bison fur flew into the swirling air, "but we couldn't do it. The animals struggled a lot. Barely got the collars on, never mind gettin' samples. An older cow with a green tag. And a younger female. I took pictures," he added. "I'll send them to you when we get home."

"The animals took off strong when we let 'em go," the son reassured me as he climbed back into the helicopter. "Last we saw of them, they were moving real normal."

The father pulled on the stick, and they rose up and headed west.

Unfortunately, "moving normal" was not the story that emerged over the next few days when the locations of the newly collared bison appeared on our computer screens. The other animals moved normally between valley-bottom meadows at their usual pace of about two kilometres a day; the two recently net-gunned animals went ballistic, travelling up dead-end canyons and bouncing off cliffs, like they were in a giant game of pinball on steroids.

Seeing that made me remorseful; the collaring had obviously been more traumatic for the bison than the

father and son had suggested. Given that they were bouncing around at over thirty kilometres per day, we were lucky
the two females didn't ricochet out of the park. It was a
week before they settled down. I thought there had to be
a better way.

One pattern I dwelled on was how differently the bison
reacted when they encountered people on foot versus on
horseback. When they saw hikers, the animals often fled.
When they saw people on horses, they usually didn't react
at all.

Some fellow park biologists had started to travel by
horse to dart and collar elk for a study on the nearby
Ya Ha Tinda Ranch. On my next visit there in the spring, I
conferred with the two Parks Canada cowboys who raised
and trained the horses for backcountry use.

"Would a similar method work with bison?" I asked. "Approaching them on horseback?"

"They're both herd animals..." one of the cowboys began cautiously. He spat out his chewing tobacco as his partner chimed in.

"Bison are different from elk. They're not as predictable." He held up both hands to his head, pointing his index fingers up. "They're more likely to lash out with those pointy horns." His colleague confirmed with a nod.

"But we'll help you if you want to try."

Unfortunately, neither of them was able to join us on our inaugural horse backcountry bison collaring trip when we set off from the Ya Ha Tinda Ranch one July morning the following summer. It was 2020. Two years had elapsed since we had released the bison from the pasture, and two more of the original collars had fallen off prematurely. That meant we were down to four functioning collars: two from the fourteen we had put on the original sixteen adults in the holding pasture, plus the two new ones. It not nearly enough to track a herd that had grown to more than sixty animals and was on the verge of splitting into smaller groups.

We set off with a focused team: Saundi and Dillon to dart animals and put on collars; Adam to ensure any curious bison stayed away; me to direct and help where needed; and Sean Elliott, an understated cowboy who ran the Parks Canada horse barn in Banff, to manage the ten horses—enough to have one for each person to ride, plus a few to carry the immobilization equipment and our food and camping gear for eight days out.

I was not used to travelling in such a large group, and as we ventured into the remote bison zone, it felt like we were invading. That first night, we set up at a patrol cabin on the edge of one of the bison's favourite meadows and rode from there for the next two days, looking for targets. We saw a big group of more than forty bison, but an ideal situation—namely, away from water or steep slopes that could be dangerous for a tranquilized animal—failed to line up. We quickly learned that in order to get close to the animals without them fleeing, we needed to approach in the open, where they could all see us. This worked best when we slowly zigzagged toward the bison at indirect angles, a trick gleaned from our stockmanship training.

Everything appeared to be in order on the third day, until Sean went out to check the horses in the morning and came back with raised eyebrows, uncharacteristically excited.

"They're here," he whispered, pointing to where the grassy meadow disappeared over a small rise. "In front of the cabin, just out of sight."

Within thirty minutes we had five horses saddled, the sedative loaded into primary and backup darts, and the oxygen tanks, radio collars, and other immobilization supplies loaded on a sixth steed, ready to go. Saundi mounted her horse, Skipper, an old veteran that never flinched when a gun was fired from his back. I handed Saundi the dart rifle once she was seated, then swung onto my own experienced horse. Adjusting the reins, I went through my mental checklist: paintball pistol loaded with chalk balls on my belt; a canister of bear spray beside it; a wrench in my back pocket to secure the radio collar; and binoculars and a rangefinder hanging around my neck. I patted my chest pocket to confirm I had the plastic waterproof box with the loaded darts.

"Ready?" I asked Saundi.

"Ready," she replied. After a quick radio check with the rest of the team, instructing them to sit tight, we squeezed our legs against the horses' bellies and were off.

The bison were visible within minutes. They looked relaxed and content and paid no attention to us as we crested the rise and came into their view. We circled the herd from upwind to be sure they knew of our presence a hundred metres away. The ones that were bedded remained bedded. Those chewing their cud kept chewing their cud. When we were fifty metres away, Saundi dismounted, and I handed her one of the darts to load. We both inhaled deeply after she remounted her horse with her loaded gun.

"Let's do this," I said.

"Let's do this," Saundi echoed, guiding her horse to fall in step beside mine.

At about forty metres, a bison finally looked up. I read the cue that we were entering its pressure zone. I slowed and shifted to the zigzag method, nearing the bison at indirect angles. The secret, we had been told at Ted Turner's ranch, was to not aim directly at the animals but instead to appear as if you intended to pass them, which the bison interpreted as non-threatening. This made for painfully slow progress, with numerous pauses and nudging reluctant horses. But it worked. Bedded animals stayed lying down, a month-old calf nursed from its mother, others calmly continued to graze. Some bison even moved toward us.

"It's like we're invisible," murmured Saundi as I grabbed the rangefinder and brought it to my eye. Number 17, the outcast, peered back at me from her usual spot near the back of the group. Her green ear tag glistened in the mid-afternoon sun.

"Thirty metres," I whispered into the radio microphone strapped to my chest. "Bison calm. We're moving in."

I had only been among such relaxed wild animals a few times before, and never while astride a horse. The bison's calmness travelled up its limbs into my legs in a universal language of warmth. Time hiccupped in that moment, and for a few suspended seconds, wild and domesticated beings inhaled each other's exhaled breaths. It was different than in the pasture; more mutual and freer, without hay or fences constraining what had developed into autonomous lives. I sat in my saddle watching them, avoiding eye contact. My chest swelled in admiration and respect. They had proven so adaptable and, with the unearthing of their ancestors' bones, were such astonishing students of time. Their presence hummed all around us. Was that energy the herd's collective mind?

Saundi snapped me back: "She's stepping broadside," she announced. I scrambled to read the distance with the rangefinder. Out of the corner of my eye, I saw Saundi drop her reins and smoothly raise the rifle to her shoulder.

"Twenty metres," I said in a hushed voice as the meaty part of Number 17's rump came into view. Saundi seized the moment. She clicked off the safety, aimed, and fired.

Havoc broke loose the moment the gun cracked. In a freak accident, the cow flicked her tail in the split second the dart flew toward her and intercepted it. We both helplessly watched as the needle passed through the shaft of her tail and spewed the drug harmlessly in a mist in the air. The dart tube, with its bright pink flight feathers attached, was lodged in her broom-like tail. As she tried to shake it off, it attracted the rest of the herd like a waving neon flag. Number 17, usually one for the sidelines, was now the centre of attention. We watched her get chased around the meadow for a few minutes before she galloped into the trees with the rest of the herd in close pursuit.

"What's going on?" Dillon asked over the radio.

"Shot fired but missed," I answered. "Everyone can stand down."

I was proud of how close we came to tranquilizing a bison from horseback, but Saundi was quiet that evening and had a sleepless night. The next morning, she saddled her horse in silence and rode off to check the nearby meadows for bison activity.

None of us thought she would discover much, so we were surprised when her voice came over the airwaves a few hours later.

"They're here!" she said excitedly. "I'm watching them from the escarpment. Number 17 and two other mothers with their calves. Six animals in all. The dart is still in Number 17's tail."

We were packed and riding in ten minutes and caught up to Saundi within the hour, but Number 17 and her companions had moved on.

"They drifted that way," Saundi said, pointing apologetically, "about half an hour ago."

We scoured the meadow, an alluvial fan the bison frequented that had been cleared of most trees by frequent floods from a nearby creek. The grassy openings followed the flow lines of shallowly buried gravel beds, separated by rows and islands of scraggly forest. We spotted fresh dung and bison tracks as we searched the area, but found no animals.

"They evaporated," I said when we reconvened after two hours of fruitless effort. Everyone looked ready to give up.

"We'll ride around one more time." I motioned for Dillon to follow me. "The rest of you, stay here and keep a lookout."

Midway through the last round of searching, my horse had perked up at a certain spot. When we returned to that same spot, we found the six bison bedded down by the river, looking like they had been resting there all along.

Dillon had taken over the rifle from Saundi, and I still had the loaded darts and all the other gear, so we were ready.

"We see 'em," I barked into the radio. "Stand by. We're going in."

The river was only fifty metres away—a significant hazard for a tranquilized animal—but the dart still lodged in Number 17's tail from the day before changed my calculus. Having her in a group of only six bison also meant other animals were less likely to bother us if we darted her.

Dillon and I zigzagged to within forty metres, and I called for the other members of the team to start toward us. As they did, we inched into that humming zone, courtesy of the invisibility afforded by our mounts. The bison, only about twenty metres away now, accepted us as if we were a couple of grazing deer. The calves continued nuzzling into their mothers' sides, sniffed, and tried to focus their glassy black eyes on us. The cows remained unconcerned, their eyes half closed.

Dillon and I were suspended in the timelessness and formlessness of that moment, feeling fully accepted after years of effort, but the ease quickly vanished. A restlessness descended over the six animals; they started to get up, yawn, stretch, and defecate. I saw Dillon lift the rifle to his shoulder as Number 17 made to follow the others. As soon as she stood, a decisive shot rang out.

This time the dart found its mark. It stuck into her thigh, but, unlike before, none of the bison ran. Number 17 took a few steps; looking confused, she turned to sniff the pink fletching that now dangled from her flank as well as her tail. The others just stood by and watched.

"Dart in," I reported on the radio, noting the time. "Come quick."

Four minutes later, she wobbled, and her front legs buckled. The rest of her collapsed as she came to rest on her side. Suddenly, we had a new problem: the other bison were standing protectively over her.

"Go on, get!" I shouted, kicking and charging my horse to within a few body lengths of the animals. They held their ground, staring at us perplexed. I backed up and charged again, this time waving and slapping my thigh as I grappled with the reins with my other hand. Dillon joined on his horse for a third and fourth charge, but they stayed put. Finally, when I shot my pistol and a chalk ball flew out and hit one of the cows, the protective brood fled in a puff of white dust.

"Keep your eyes open," I said as the rest of the team arrived and everyone dismounted. "The other animals might come back."

Thankfully, we were left alone to focus on our assigned tasks: tie the horses to nearby trees; lube and cover the bison's eyes; reposition her to prevent aspiration; monitor vital signs; administer oxygen; depressurize and remove the darts; attach a radio collar; and take blood to later test for diseases. I read the checklist back to Saundi when we were finished, to double-check our steps.

"All clear?" she asked, drawing up the reversal drug into the syringe.

"All clear," I answered after we had packed up and stood to the side.

Three minutes after Saundi plunged the needle with the reversal drug through Number 17's thick hide, the bison's muscles twitched beneath the bristly fur of her hind end. Seconds later she stood up and charged away. She stopped long enough to sniff and pick up the scent of her calf at the edge of the meadow, then followed it into the trees.

"SO, LET ME get this straight," Sean clarified, passing out drinks once we were back at the cabin. "That was the first horse-mounted capture of a wild bison that's happened in over a hundred years?"

I looked at Saundi, Dillon, and Adam before I answered. Rosy-cheeked and smiling, they beamed with pride.

"As far as I know," I replied, and we all toasted each other.

≡ **17** ≡

LEVERAGE
SUCCESS

REINTRODUCING BISON TO Banff National Park was described as a five-year *pilot* project when it began in 2017.

"Let's see if it works," Parks Canada had said, acknowledging the many uncertainties the project posed and the previous failure in Jasper. "If it doesn't work, we'll round up the animals, remove the drift fences, and pack up."

Looking back in 2023, the Banff reintroduction was a huge success. The sixteen original animals had grown to more than a hundred; none had become sick or diseased; there were no obvious wolf chases outside the park; no negative animal interactions with people; no property damage; and—except for a few wayward bulls—the animals mostly stayed inside the park. Local Indigenous Nations were inspired; the project validated an ancient prophecy that wild buffalo would someday come back via the sky and mountains.

However, there were many warnings from the bison that tempered that optimism. As the herd grew and the number of animals increased, they split into subherds. Some subherds lacked collared individuals and could not be tracked. Responding to every excursion out of the park became impossible. More animals meant more breaches and damage to the drift fences. Perhaps their ancestors were speaking to them; 98 percent of the bison's movements pointed northeast, to the Ya Ha Tinda area outside the park.

The success of our pilot bison reintroduction could have been leveraged by Parks Canada to invite Alberta to co-manage bison in the only way that is sustainable into the future: in a transboundary area that would support about a thousand animals. Biologists agree that this number is enough to insulate the herd from the threat of going extinct from fires, floods, disease outbreaks, or inbreeding. Unlike the national park, such a transboundary area would have enough motorized access to make it possible to manage the herd's growth with Indigenous and non-Indigenous hunting—a concept we had discussed since the beginning of the project, but that was not logistically possible within the park. But instead of leveraging the project's success with the government of Alberta, Parks Canada did not maintain their support for the Banff bison, and our budget dried up. Those of us who had dedicated many years of hard work to bringing the bison back felt it like a kick in the guts.

Looking back, I suspect Parks Canada's waning support was not intentional, but the result of numerous changes that created a perfect storm of bureaucratic obstacles. The first change was yet another restructuring at the national office, where a new team of employees was hastily assembled and cut projects they knew little about. Advocates for the park, like my boss, Bill Hunt, and his boss, the superintendent of Banff National Park, who would normally counter such national office oversights, both retired. Their replacements lacked the experience to deal with such matters and were unable to help.

Meanwhile, my own capacity had been severely curtailed. In October 2021, I had a serious fall from a tree, which broke my back and neck in fourteen places. The fall likely triggered multiple system atrophy, a rare condition that attacks the muscles via the nerves. After months of tests, I was delivered a terminal diagnosis. I was fifty-five years old, and the body that had served me well for my whole life was suddenly, prematurely shutting down.

Worried about the lack of future support for the project, Adam had left two years earlier to take a more secure job elsewhere. Saundi and Pete now followed. As my disease rapidly progressed, I became useless, tipping into walls at the office and too weak to mount a horse. I submitted the paperwork for a medical retirement. Amid all this, the national office thankfully recognized that the Banff bison required some ongoing attention and, at the last minute, accessed some emergency funds to keep Dillon employed to manage the project. Although the commitment lacked adequate funding for herding responses, or for enough radio collars to monitor the animals, Dillon hesitatingly agreed.

I was slated to leave my job at the end of March 2024, a time of year known as March Madness, when managers spend more freely because of the weird government policy that forbids carry-over of any unspent funds between fiscal years. Despite the impending financial crunch, I had a few thousand dollars left over in the 2023/24 bison budget. It was time to rent a helicopter and figure out how many animals we had on the ground.

Knowing this number was important, especially because discussions had started with Indigenous groups about ceremonial bison hunts. I handed Dillon a folder describing various census techniques and left him to plan it.

My declining health dictated that that flight would be the last time I would see the bison. I was not in pain but had lost most feeling in my left leg and required a ski pole

to hobble to the helipad where I greeted Dillon, the pilot, and a biologist from the government of Alberta, whom we had invited to join us. If any of them noticed how awkwardly I clambered into the front passenger seat, they didn't let on. They waited patiently as I sat down and lifted my feet onto the floor of the cockpit with my hands.

It was a glorious day for a flight—cold, clear, and calm—and I drank in every detail as if I were in a helicopter for the first time. I never tired of that trip, of leaving behind the black lines of asphalt and glinting buildings and cars of the busy Bow Valley and swinging north into an expanse of whitewashed valleys and peaks. All that sullied that pristine blanket of snow was some snowy rubble from a few winter avalanches and the posthole prints from some intrepid bands of bighorn sheep.

That changed twenty minutes later when we crested a ridge and flew over the Panther Valley.

"Whoa." I signalled for the pilot to slow down as I fished my GPS receiver from my pocket. What had been a smooth blanket of snow was now a maze of pockmarks and divots. "There are tracks and feeding craters everywhere!" blurted the biologist.

It had been months since any significant snowfall, and from that height a calligraphy of curved lines and punctuation marks revealed a story of animals that had explored every nook and cranny of the valley all winter. The signatures of their activity were astounding. The bison had returned!

"Remember, we're looking for animals," I reminded everyone as my GPS powered up. When it did, I directed the pilot to the start point of the census that Dillon had designed, which was located only a few hundred metres away.

"Start here!" I held out a hand in the direction the pilot was to follow and asked him to fly us steadily at fifty kilometres per hour, five hundred metres off the ground. Such a consistent approach was necessary for Dillon or others to compare census results in future years. Dillon had derived the grid we followed from collar locations, which assumed we would not spot every animal. Probability equations applied to the number of bison we counted would provide the final estimate. But that was tomorrow's work—right now, we just had to look for animals.

Bison are easier to spot from the air when there is snow on the ground. After no more than a kilometre, I saw our first group—thirty-four animals stretched out in a line on a grassy ridge. Even though the ridgetop was blown clear of snow and there was no white for the bison's dark-brown bodies to contrast against, their shadows on the slope behind gave them away. As the pilot hovered, I marked the location with my GPS, adjusted my binoculars, readied my camera, then slid the passenger window open to snap a few photos.

"Three mature bulls," I counted as I held the binoculars to my eyes. "Fourteen calves."

"Looks like two have radio collars," chimed in the biologist. "Zero percent snow. Twenty percent tree cover." Dillon furiously scribbled down the data.

"We'll confirm which animals those collars are on later from the photos," he said, thinking ahead to the probability calculations when we returned to the office. Each collar had a vinyl placard with a unique letter bolted to the side that was visible in the photographs.

No sooner had we finished with one group than our biologist friend saw another.

"Up there," he said, directing our gaze to the grassy slopes that unfolded in a chaos of tilted plains, knolls, and short, narrow valleys. Two big bulls, their curly foreheads covered with snow, lay like brown beacons in the deep drifts near the top of the slope. We flew overhead, marked their location, then returned to the grid.

"There!" Dillon exclaimed as movement near the river caught his eye. Veering from the grid again, we flew down and counted: four cows, two collars, no calves.

For the next hour, we continued our search over the Panther Valley. We counted ten groups in total, including

a lone bull revisiting the old holding pasture. Dillon and I were both astounded.

"What we're seeing isn't the story told by the collar downloads this winter," Dillon said. "The animals and their activity are much more widespread. Just goes to show," he joked, "you can't believe everything you see on a phone or computer screen."

The tenth group—six cows and three calves—wallowed through chest-deep snow in the high meadows where the whole herd had spent their first winter. It was an elevated, flat area between the Panther and Red Deer watersheds, and before we proceeded to search the latter, I motioned for the pilot to pause.

"How's your fuel?"

He glanced at his gauges. "Good," he nodded. "We have plenty."

"Okay," I said, seized by an idea. "Take us high." I pointed up. "Above the ridges."

He obliged. Soon the whole bison zone lay below us, splayed like a giant, three-dimensional carpet. I had the pilot turn the helicopter so the biologist and I both looked east, out the windows on the left side.

I am unsure if it was my frustration with the project or my grappling with the bad news of my diagnosis, but I did not waste time getting to the point. I laid it out simply for the Alberta biologist, passing along the lessons the bison had shared with us about what they needed to survive. I cleared my throat.

"See all that grass out there?" I pointed to the Ya Ha Tinda area beyond the park to the east, an island of gold beyond the green forest below us.

"Uh-huh," he answered.

"It's outside the park, and *that* is where the bison want to go." I thought of all the old bison wallows I had seen there with Wes Olson and the plethora of old bones that must be in the ground. "It's calling to them," I added. I then described the recent analysis that showed 98 percent of the reintroduced bison's movements were oriented there.

"But they can't go there because of the Alberta government's no-wild-plains-bison policy," I said bluntly. I pointed to where the cutline of the park boundary separated treed mountain valleys from all that grass. "It's a lot of work to keep the animals in the park." I described the ongoing tasks of checking and maintaining drift fences, herding wayward animals, and monitoring their movements 24-7. Just saying it exhausted me, never mind the effort. "It's going to get tougher as the population grows," I added, citing the bison herd's growth rate of over 30 percent per year. I traced an upward curve in the air with my hand. "Do the math yourself. It gets scary. Logarithmic. One thousand bison by 2030." I let him absorb the numbers before I guided him toward the most obvious solution.

"The reintroduction's been a success. So much so that we need to reintroduce hunting. Soon."

"What?" He struggled to wrap his head around the irony of the situation—hunting animals we had just worked so hard to reintroduce.

"It's the only way bison can stay here." I recalled Wes's impression that the best bison habitat existed outside versus inside the park, and how the animals showed that through their movements. "Other than at the Ya Ha Tinda

Ranch, there are no conflicting land uses in the twenty-kilometre-wide band of public land around the park," I said. "Not with cattle or any other industries." I explained how a regulated bison hunt would help keep bison numbers in check and discourage them from ranging beyond. "That strategy worked in the Yukon," I said, describing how hunting limited the range of a reintroduced wood bison herd there. "It could work here."

I pointed to the grey line of a gravel road threading through the clear-cuts into the area from the distant prairies. "Some motorized access, but not too much. A way for people to get hundreds of pounds of meat from each hunted bison out. Of course, Indigenous hunters would have priority," I said, "but there would be enough for non-Indigenous hunters too." I drew his attention to the hazy Calgary skyline in the distance—a city of 1.6 million people—and referred to the popularity of the wood bison hunts in northern Alberta. "Uptake from hunters here won't be a problem."

"Definitely not," he chuckled.

I reminded him how wildlife ranging in the area was already managed by Alberta, and how the last mountain ridge bounded its extent naturally. "A gift of topography," I said, pointing to the rocky ridge that could block bison from spreading east into the foothills. From that elevation, it looked like the edge of a giant corral built by the gods. "A natural boundary to keep bison away from cattle."

Leaving our biologist friend to digest these thoughts, I asked the pilot to resume our grid. The full northward extent of the Rockies' Front Ranges filled the windscreen as he turned us, and I considered sharing an even bigger vision: the Clearwater, Ram, North Saskatchewan, Cline, Brazeau, and Southesk valleys were all visible grooves in the curvature of the Earth, each offering grassy and alpine meadows for bison. Each of their headwaters were already protected in Alberta's Siffleur, White Goat, and Willmore wilderness areas and in Jasper National Park. Between the two governments, Alberta and Canada managed more potential bison habitat than anyone on the planet. I imagined a future where more than a thousand of the animals could range below us, enough to survive the fires, floods, inbreeding, disease, and other natural disasters that threaten other isolated populations with extinction. Doing so would not just be a legacy for Albertans and Canadians; it would also contribute to the global conservation of a threatened species.

I did not mention this grand vision to the Alberta biologist. I didn't want to take his focus from the next step: co-operating with Parks Canada on a transboundary bison zone in the Red Deer and Panther valleys, where Indigenous Nations and the public would help manage expanding bison numbers and their range.

We searched the Red Deer Valley portion of our grid, but spotted only one more big group—forty animals. They lay in a tight-knit patch among the charred stumps of a decade-old burn. A handful of late-born calves wove through the brown dots of bedded mothers and black dashes of fallen timber like red fireflies. We counted and photographed them, then finished the census with a total of over ninety animals seen in eleven groups.

As we flew back to Banff, I thought back on all the lessons the bison had taught us, and whether I had been constructive in how I'd passed them on to the biologist. He was not a decision-maker in the Alberta government, but he frequently talked to those who were. How he would communicate the bison's teachings remained to be seen. In a wave of the acceptance that accompanies someone who knows he is dying, I was content knowing I had done my best. Other, bigger forces were at work—the bison were advocating for their own expansion. It was as prophesized: Bison had come from the sky, and now they were repopulating from the mountains.

"The animals are going to keep asking for what they need," I said as the Bow Valley came into view and the pilot prepared to land. "It's not a question of *if* others get buffaloed, it's *when*."

EPILOGUE

THE FIRST INDIGENOUS bison hunt in more than 150 years occurred in what is now Banff National Park as I finished this book. The numbers were telling: Of the nine nations invited to hunt bison in the backcountry of the park, six took part, and of those, three were successful in their hunt. Given the remoteness of the location, extracting the meat was a struggle for those who did connect with an animal.

Despite the challenges, the inaugural year of the hunt was labelled a success. I agree, insofar as it is the most tactile example of Indigenous reconciliation I know about; we had reintroduced not only a species but also the human connection to it.

However, the response of the animals was concerning. Still able to download the collar locations, I watched them surge from where they were hunted in their favourite meadows to outside the park. It reminded me of the pattern after we'd net-gunned two of the animals to collar them. Dillon and others had to herd them back into the park and watched with concern as they continued surging, bouncing off a drift fence along the park boundary.

It was yet another lesson from the bison. Until then, the national park had been a safe refuge for them. If they are regularly pursued by hunters, will they still view it that way? Will they stay within the park boundaries? Only time, and the bison themselves, will tell.

ACKNOWLEDGEMENTS

THANKS TO THE following people for reviewing and commenting on earlier versions of the manuscript: Leanne Allison, Erica and Sigrid Heuer, Miriam Körner, Jane Kubke, Quincy Miller, Joseph Obad, Wes Olson, Saundi Stevens, Dillon Watt, Alex Taylor, Jesse Whittington, Cliff and Peter White, and Adam Zier-Vogel. Double thanks to Cliff White for planting seeds in the Parks Canada bureaucracy for the bison reintroduction to happen. Harvey Locke and Marie-Eve Marchand germinated those seeds through their Bison Belong work with Banff's Luxton Historical Foundation. I also appreciate the early work of Parks Canada wildlife veterinarian Todd Shury, and biologists David Gummer, Tom Hurd (retired), and Brian Low (retired) for shaping and promoting the bison reintroduction idea. Thanks to the staff and horses at the Ya Ha Tinda Ranch and Don Gorrie and the rest of the Banff Park trail crew for their help implementing the project. Caroline Hedin communicated it to a wide and diverse audience in the early years when we had money to hire her. Thanks to Leroy Little Bear and Amethyst First Rider of the Kainai Nation/Blood Tribe, and Bill Snow of the

Îyârhe Nakoda Nation, for your friendship and collabora-
tion. You all helped me digest the bison lessons for myself.

My mind grew fuzzy as I approached the end of this
manuscript and my life. A friend and past freelance edi-
tor, Helen Rolfe, stepped in to help at just the right time.
Through her suggestions and our talks, I saw how what
I began five months earlier had become muddled. She
helped me hone it into the simple lessons the bison had
shared with us.

The terminal nature of multiple system atrophy meant
I spent much of my final six months in the ten-by-ten-foot

writing shack in my backyard or at the more spacious West Coast retreat of my friends Don and Betty-Anne Graves, who graciously offered it as a quiet place to write. Thanks to my wife, Leanne Allison, my son, Zev, and my many friends who recognized the importance of that work to me as my time narrowed. Thanks to Rob Sanders, Paula Ayer, Jen Gauthier, and the rest of the team at Greystone who agreed to publish and distribute this book.

FURTHER RESOURCES

FOR MORE INFORMATION about the bison reintroduction to Banff National Park, visit:

parks.canada.ca/pn-np/ab/banff/info/gestion-management/bison

This site, developed for the project, includes numerous photographs and videos, as well as a blog with dozens of entries written in the early years of the reintroduction.

For more information about Indigenous efforts to bring back buffalo, visit:

buffalotreaty.com
buffalorelations.land

For a thorough natural history of bison, see the 2022 book *The Ecological Buffalo: On the Trail of a Keystone Species*, by Wes Olson and Johane Janelle, published by the University of Regina Press.

DAVID SUZUKI INSTITUTE